HOMINESCENCE

MICHEL SERRES

Hominescence

Translated by Randolph Burks

HOMINESCENCE

Michel Serres
Académie française

BLOOMSBURY ACADEMIC
LONDON • NEW YORK • OXFORD • NEW DELHI • SYDNEY

BLOOMSBURY ACADEMIC
Bloomsbury Publishing Plc
50 Bedford Square, London, WC1B 3DP, UK
1385 Broadway, New York, NY 10018, USA

BLOOMSBURY, BLOOMSBURY ACADEMIC and the Diana logo are trademarks
of Bloomsbury Publishing Plc

First published in France © Editions Le Pommier 2001

Copyright to the English translation © Bloomsbury Publishing Plc 2019

Cover image: *'luminescence 41'* © michaeltaylorphoto.com

All rights reserved. No part of this publication may be reproduced or
transmitted in any form or by any means, electronic or mechanical, including
photocopying, recording, or any information storage or retrieval system,
without prior permission in writing from the publishers.

Bloomsbury Publishing Plc does not have any control over, or responsibility
for, any third-party websites referred to or in this book. All internet addresses
given in this book were correct at the time of going to press. The author and
publisher regret any inconvenience caused if addresses have changed or sites
have ceased to exist, but can accept no responsibility for any such changes.

A catalogue record for this book is available from the British Library.

A catalog record for this book is available from the Library of Congress.

ISBN:	HB:	978-1-4742-4786-3
	PB:	978-1-4742-4704-7
	ePDF:	978-1-4742-4705-4
	eBook:	978-1-4742-4706-1

Typeset by Integra Software Services Pvt. Ltd.
Printed and bound in Great Britain

To find out more about our authors and books visit www.bloomsbury.com
and sign up for our newsletters.

For Lise and Pauline,
Marie, Magali and Victoire,
Jean-Baptiste, Claire and Thérèse,
Josué, Raphaël and Irène,
in the past called the Captain.

CONTENTS

Deaths 1

PART ONE THE BODY 13

1 How Our Body Changed 15

2 The First Loop of Hominescence 39

3 Three Global Houses 53

4 The Greatest Contemporary Discovery 59

5 *Ego*: Who Signs These Pages? 67

PART TWO THE WORLD 69

6 The Greatest Contemporary Event 71

7 Old and New Common Houses 83

8 The Evolutionary House 129

9 The Second Loop of Hominescence 139

10 Who, *Ego*? 149

PART THREE THE OTHERS 157

11 The Event of Communication 159

12 Contemporary Humanity 175

13 The End of Networks: The Universal House 191

14 The Third Loop of Hominescence 211

15 The Others and the Death of the *Ego* 221

Peace 231

Notes 267

DEATHS

Eulogy

We no doubt became the humans we are from having learned – will we ever know how? – that we were going to die. We find the only honest remains from prehistory and high antiquity in tombs, bones accompanied with objects. Animals have neither objects nor death. So this dreaded end belongs exclusively to us twice over: insofar as we are humans, insofar we are singular individuals; it awaits us and reaches us in our generic definition and our singular solitude.

But, by ending up destroying our lives, death constructs them: without the stiff cadaver it leaves behind, without the sex it was long believed to imply or the irreversible time it brings about, would we have ever painted the walls of caves, lit fires, sung within the lacework of language, danced for the gods, observed the stars, demonstrated geometrical theorems, loved our companions, educated children, lastly lived in society? In *The Antique City*, Fustel de Coulanges demonstrated that the worship of dead ancestors predominated before the classical era; houses had tombs for foundations, and metropolises began as necropolises. I tried, in *Statues*, to generalize his analysis, limited as it was to the Greco-Latin area, by giving it anthropological value. Behind our backs, death and the weaknesses issuing from its sentence engendered human civilizations. The questions 'Am I going towards death?' or 'Am I delivering myself from it?' construct meaning.

What are we now risking by excluding it from our thoughts, from our customs, from our personal behaviour, from our collective rites? Non-meaning, the non-human? Worse, what are we risking by attempting to live with almost no risk since she for whom or that for which I would give my life holds its meaning? Life and value aren't equivalent: life measures value and measures it only because death keeps them in finitude.

The West's originality: Its antiquities

Civilizations die, too, in the same way individuals do and in the same fashion, as certain as it is unforeseen. We who are witnessing the end of agrarian cultures, which arose during the Neolithic, the disappearance of ancient languages, the murder of European taste, the sudden vanishing of political systems whose durability was predicted, not long ago, by so many militants, we have long known that civilizations are, like us, mortal. The originality of ours consists in the fact that an antiquity precedes and founds it.

For the era that just celebrated its second millennium began on the ruins of Rome, an earthly city its contemporaries had believed to be immortal. Yes, the distinctive feature of the civilization called Western comes from the fact that it is erected on the disappearance, but at the same time on the retention of the ancient civilization it denies. The event of the resurrection of Christ, which Saint Paul said founded Christianity, signifies, from this perspective, that contrary to the ancient body and city, our bodies and cities turn their backs on death; not only would the holy women's spices and the strips of linen in the tomb no longer be used for Christ's mummification, but Saint Augustine, in *The City of God*, an innovatory text of the new era, generalized this idea: the time of the new City starts from the end of the earthly Rome, a death integrating the deaths of several antiquities – Greek, Egyptian, Latin … – all of them drugged with polytheism.

So tomorrow, in three minutes or a few years – we don't know when – we are going to die, from illness, from an accident or from exhaustion. Likewise, we don't know when the largest current power in the world, obese with money, will collapse, perhaps without making much of a racket: next year, in six months, in a hundred years? But we cannot help but to have learned that these two rendings and these two not-knowings found our knowledge and practices. If we ever forgot these conditions of our exemplary arts and of our excellent behaviour, tomorrow we would dance before our cathedrals the way monkeys chatter in the jungle-invaded temples of the Yucatan and Angkor. From this dynamic and vital, individual and collective function of death, meaning shoots forth.

The third and new death: A global one

So up until a recent date, we underwent two deaths: the one that seems to be the only interesting and original one, our own death or the one of she that we love; yet there's nothing more ordinary and commonly shared.

We knew as well that entire cultures had disappeared: frequent, common, ordinary and blind like the first one, this death strikes and will strike in just as unpredictable a manner.

But a third death, unknown to the human race all the way up until halfway through the last century, exactly on 6 August 1945, marks one of the two or three great innovations of the epoch that is coming to an end, an epoch in which we even risked experiencing it full scale: the global death of humanity. On two mornings of wrath in which two atomic bombs conceived and constructed in the United States exploded, in Japan, over Hiroshima and Nagasaki, my generation was the first in History to learn that the human species was now risking dying out. What are we to think about the scientists who didn't hesitate, in the following decades, to supply the militaries and politicians of the most powerful countries, but also other nations, with increasingly destructive thermonuclear weapons? That the ethics or deontology, that the law, that the humanities they had learned no longer fitted the means they were providing to the decision-makers of the day. Starting from the same date, this observation was to be non-stop: hampered in its past, philosophy no longer understands the new order and no longer plans to build the house of the future generations. While technicians and scientists are causing a new world to be born, philosophy is thinking as though it were a question of the old one. Ever since Nagasaki and Hiroshima, it has already been necessary to change philosophies.

Two modalities

This global death, an original and for once authentically common one, comes to us in two modalities. The first one, undergone, can ensue from some chance natural event; reviewing the formation of our Earth, we discover traces of accidents of this type stemming from volcanism or meteors. For almost every living species disappeared 550 million years ago, then 440, 370, 250, 210 and 65 million years ago …, global and semi-periodical deaths whose occurrence, each time, reoriented evolution. A worry: the interval separating us from the most recent accident has lasted longer than one of the preceding intervals. Once again, we were born from these very deaths that didn't depend on us.

The second modality, in some way intentional, could ensue from actions that, on the contrary, really do depend on us. So the collective death haunting us comes in at least three forms: through our power of nuclear fire, first, were we to make war. Through our industrial pollution next, in the midst of peace; we fear and accelerate global transformations and, in particular, the disappearance of certain species without knowing just how far these changes or eradications will extend. Through our cruelty towards our own race, lastly, since the West puts the third and fourth worlds to death coldly, for money. Faced with these three responsibilities, how can we reorient our enterprises and perhaps our time?

The new and fourth death: A local one

After Hiroshima, the collective and global atomic terror, threatening the human race, rose above the personal and cultural deaths, common tragedies lamented by the traditional Humanities. After that, we caught a glimpse of the possible eradication of all the species, announcing our possible solitude, lastly the murder of our own brothers.

Added to this, this very morning, is a strange and fourth death, itself coded: for a tiny signal decides the elementary suicide first of the cell, then of a function, lastly of the organism. Scientists call it apoptosis, from the name the Greeks gave the fall of leaves in autumn. Will we master death by decoding this call? Who can predict this? For we discover, at the same time, an unstable equilibrium between the direct death brought about by this signal and the indirect one in which cancerous cells irrepressibly proliferate because, deaf to the call of apoptosis, they refuse to self-destruct; by means of this tumour, the organism dies from life. That is, again, the double meaning of death, and here, for the first time, is stated the double value of communication, common to every messaging system: like tongues in Aesop the fabulist, every channel can become the best and the worst, as we shall see in abundance. Here, apoptosis destroys the body but constructs it as well by sculpting the embryo and eliminating superfluous cells from it. Like the preceding ones, this death therefore results in two values: negative and positive, deleterious and constructive. This fourth death announces the profound unity of a book attempting to assess the scope of the current mutations by which the body and communications are bifurcating. Apoptosis touches all these mutations and unites them, for it destroys and constructs the organism via signals. Although distressing and truly contemporary, this cellular and messengerly death presents itself to us as endowed with the same meaning as the other ones. It causes us to disappear, of course, but shapes our formation, our muscles and our nerves, and so determines our sensory and motor performances, in brief, kills us but invents our life as well, even more concretely than the other ones.

Sung by the arts, meditated by religions and philosophies, the two traditional deaths strike individual bodies and local groups. The two new deaths frame the two first ones, symmetrically, the one in globality (human race, living species, entire planet), the other in the minuteness of the elementary cell and the tiny signals that constitute it. We no longer live the same deaths as those we have undergone ever since our origin, especially as we will perhaps master the new ones. Is there an event more decisive than this one in the process that made us into the humans that we are? Too preoccupied with culture no doubt, philosophy kept silent during the recent moments when these two events appeared, events overturning what remained of nature, from its microscopic birth to its entirety. Different deaths, unforeseen time, new humanity.

Therefore, whether individual or collective, elementary or global, voluntary or involuntary, these four deaths together show a subtle equilibrium, one incessantly pushed to the point of instability, between a power of vital emergence and another power that cuts and eliminates. Once again, we can't think life, and humans in particular, without death, nor can we think the absurdity of the latter without giving meaning to the former. The energy governing the formation of the one goes hand in glove with the other. This couple works to bring about time. Thanks to oxygen, I breathe, and its rust destroys me. What kills me comforts me.

Two apparent exceptions, only one of them real

Global and never coming back, does the eradication of the living species under our responsibility form a first exception to this optimistic tension? What would we gain from remaining alone in the world? But we don't really know how to define this disappearance, for we are still and always discovering species, especially insects, arthropods and single-celled organisms …, and we assess poorly the species that remain to be counted. In addition, at the same time we are evoking this eradication, we are beginning to decode the library of destroyed genes and to dream of regenerating species or even to invent new ones. The mingling of our excellent knowledge and our extreme violence clearly lies at an unfathomable depth.

The second exception has long choked our throats up with an inexpressible anguish. I don't know of any optimistic side to this abomination: the wealthy of the West, contingently my brothers, allow more than three quarters of humanity to die, who thereby become my essential brothers. Who do I call my neighbour?[1] The one who risks a premature death, especially under our responsibility. Of course, another humankind is being born today, but in the midst of bodies deceased by a global and conscious crime against humanity. From the lifting of this negative and from this alone will this birth occur. False gods, we will only become humans again by abolishing this death penalty pronounced against mortals, the only ones to remain human in the essential sense.

Life and mind: Two ancient immortalities

Our forebears already invented two answers to the two first deaths. Those who didn't limit their efforts to training, taming and subduing certain animals but who succeeded in domesticating other animals mastered, more than their bodies, their lineages. For ever since the beginnings of livestock farming, no ewe, cow or bull has ever been seen to return into the woods, to the life of the wild, forgetful of our lessons. They would die from it. The first person who, having enclosed a piece of land, took it into his head to say 'this is mine' invented livestock farming and agriculture, whose practitioners know, better than philosophers, that domesticated plant and animal species weaken to the point where they must be defended, at least by an enclosure, against the aggressions of the wild species, more powerful than them. On this condition, which all by itself defines rusticity, plants and animals blossom and endlessly reproduce. *Detachment* sang the praises of these genius forebears, whose names have been erased, going as far as saying that they invented immortality. In fact, even though we no longer know where wheat came from, we have never lost it; wheat, cattle and sheep have uninterruptedly supplied us with food ever since this discovery, one easily found by each generation.

We know little history since even our children are wildly mistaken about the wars we went through in our youth; we have forgotten the memory of gods, kings, battles, customs, works and days; the only things we have preserved, certain and stable, are lineages subject in the past to domestication and mathematics submitting to demonstration. Truthful history frequents the peasantry and the rigorous sciences, rare sites where sublime geniuses invented this type of immortality; like wind, flesh and grass, the rest are tattered with forgetfulness and lie. In company with Pythagoras's statements, guinea fowl and wheat accompany our time with their faithful invariance across variations and selections.

A new immortality?

Between 1970 and 1980, between 'The Thanatocracy', whose poignant accents announced the advent of a new death, the death, a global one, of the human race beneath the flash of nuclear bombs, and *Detachment*, which sang of the stable lineage of the domestic species we inherited, biochemists manipulated the genome and therefore changed livestock farming and agriculture: Are we heading towards a second type of 'immortality'?

Evolution results from selections and mutations. Darwin drew selection from agrarian practices and thus accounted for the survivors, sometimes crossbred, always fitter; mutating creates newnesses.[2] Thus, through mutation, the new appears, the best adapted of which, through selection, subsists. Bodies, species, lineages appear when their genomes transform, as we now know; it takes chance and time for this. Human history doubtless would not even have begun if contingent conditions hadn't caused our own genome to bifurcate lightly and if we hadn't subsequently weighed heavily upon the evolution of the living things around us.

When our genius forefathers or foremothers 'invented' the sheep or corn starting from teosinte, they imitated nature by selecting these species through crossbreeding, through elimination and therefore fabricated phenotypically modified organisms or PMOs, a possible name for domestic species. But there was nothing they could do regarding their potential mutations, hidden in the dwarf depths of their germ cells. In attaining this and manipulating genes, we are intervening in mutation, therefore minimizing chance, accelerating time and inventing genetically modified organisms (GMOs), and perhaps species; 'immortality' has returned here since this new genome will be transmitted without a flaw. In dominating selection, these ancestors contributed towards the historical originality of human time; in mastering mutation, are we opening up a duration connected to the duration of evolution? A new time for different lives?

Evolution reconnected to history

Better, through varied selections, duration preserved wheat and dogs as relatively invariant: the same species under a number of aspects. In a certain way, we cadence time through their variations. How delightful our memory exercises would become if we taught our children the epoch of corn or sheep as divided into eras named for their various varieties! During those times, we would say, the bigarreau cherry, the beurre Hardy pear or the comice pear appeared. The *prairial* and *floréal, messidor* and *fructidor* of the French Revolutionary Calendar come fairly close to these exquisite announcements, far from the crimes of Julius and the tyranny of Augustus those hungering for cadavers force us to celebrate in August and July. For without the era of the potato, famine would have depopulated Ireland and Germany even more. Agricultural selections concerned women and men more than wars and reigns. Such selections cause them to live, the latter kill them. Our annals show that we delight less in dining than in spilt blood.

Time lasts and changes with the selections. A few decades ago, biotechnologies made that particular time marginal and slow: these biotechnologies therefore substituted mutations for selections in the archaic and sublime act of the inventors of 'immortality'. We certainly didn't stop selections, but mutations initiated and commanded them. We produce a new living thing with a lightning-fast master stroke instead of waiting for it with a throw of a dice over an unpredictably long time and selecting it with patience. A new time appears at the same time as new living things do. Different deaths, different lives: a new humanity, a different history. Will it be necessary to change this ancient word in order to describe our new time?

Nothing surpasses the importance of this bifurcation in which evolution reappears so as to plunge, anew, into history, whereas our history, up until yesterday, tore us away from it. For from our beginnings, our technologies have defended our bodies and, by protecting them more and more powerfully – via the new environment of these technologies – against natural selection, have distanced us from it. The demographic explosion gives a new proof of this today. In hominization, history dissociates itself more and more from evolutionary selection. But if we now succeed in dominating not only selection but mutations and in choosing them, we directly attain the very source of this evolution and so form a new confluence with it. The human adventure used to unfold outside the ordinary time of life; by mastering this time, this adventure in a way returns to it. So a temporal rupture without any equivalent appears.

Evolution sculpts the bodies of living things by means of death. Should death change, how would humans and their bodies not transform? Mutation causes new living things to emerge. Should they appear, how would our time not bifurcate? Should death, the body and life take on another face, how would humanity itself, under this evolutionary thrust, not change?

These unexpected deaths and these new reproductions give rise to a wind of 'immortality'. A new utopia sweeps us along. New, really? It already inspired the wanderings of Gilgamesh, our more recent ancestor, and, further back, the women and men who domesticated the ox and planted wheat. Do you condemn utopias? Have we ever built a future without them? At least this particular utopia doesn't do anyone any harm since it defies death. I therefore stand by what I said: yes, the new deaths join with dreams of 'immortality', dreams as new as they are ancient. Will my grandchildren grumble about living healthy up to a hundred and twenty years? With highs and lows, woundings and curings, an unanalysable mixture of the unpredictable and the rational, doesn't our chaotic and contingent adventure deliver us, and with as much difficulty as you might like, from Necessity?

Hominescence

Begun thousands of years ago in South America and the Middle East with the domestication of rare vegetables and animals, the age of agriculture and livestock farming is therefore bifurcating today with the invention of genetic engineering. We used to intervene in phenotypes; now we manipulate their genomes. The same intention, various actions and on another scale, towards a different direction: thus we are living through the end of one epoch and entering into another.

But what precisely is it we are calling new? I have just given some major examples of it: two deaths, literally unheard of, living things that were inconceivable a few decades ago, the emergence of a new time. Let's therefore call events new that end an epoch while referring to the circumstances that opened it up and from which these events strongly bifurcate. Biotechnologies change our relation to living things and to their duration; the atomic bomb and the apoptosis signal change our multi-millennial relations to death. All the more decisive because the epochs concerned sometimes stretch all the way to forgotten origins, these newnesses have multiplied for a half a century in order to reach a number, a diversity as well as a size that are critical enough to have repercussions on men and women themselves and produce, on balance, a kind of hominization. This book has as its object these events and their consequences, the emergence of links without any known equal to the body, to the world and to others.

Humanity today therefore seems to me to be crossing a stage amid the long duration of its contingent destiny. At the end of my life, the women, men and children with whom I live, work and think no longer maintain the same relation with the world, themselves, their bodies and others as the one maintained by their predecessors from before the last world war. I was lucky: my existence saw the human condition transform. I can say how and why. I don't yet know towards what.

Begun in silence millions of years ago, recently subject to a sudden and rapid bifurcation, our future vibrates or beats between several possibilities whose limits hesitate, as always in the process of hominization, between deliverance and catastrophe. We are constantly mixing an extreme violence with a rare wisdom. Up to now, we have benefitted from a thousand bits of luck since, with benevolent strokes of genius, our species, contingent like every species, has survived its depredations, plunderings and squanderings, hatreds and intraspecific wars.

But we had doubtless never disposed of means as effective and universal to change the world and ourselves, the air, polluted or clean, the land, enrichable or desertified, water, potable or poisonous, fire, energetic or destructive, the global climate, our inert and living environment, our individual bodies, the living species in their entirety, the function of descent, the occupation of the land and space, our relations and our collectivities, the life or death of languages and cultures, the status and continuation of the sciences, cognition in general, pedagogy and the struggle against ignorance. Each of these things and all of them taken together now depend on us, commonly speaking. Compared to our old powers, the ones we have just acquired have rapidly changed scales: we recently went from the local to the global without any conceptual or practical mastery of this latter.

These globalities have just taken on another face, one that's practical, concrete and quasi-close at hand. Everything depends on us. And through new and unexpected loops, we ourselves end up depending on the things that depend globally on us. Here, risks and chances grow as fast as our omnipotence. Behold here, in sum, our newness: it forms a sum, and we encounter it everywhere. Since this has never happened to us before, we don't know what we are to do with all these powers; the philosophy they require is forever not being born, hesitates, vibrates, trembles, flickers.

Just as with luminescence or incandescence, a light increases or decreases by flashes and occultations, a light whose intensity hides and shows itself by trembling to begin, even though ceaselessly ready to go out; just as adolescence or senescence advance towards adulthood or out-and-out old age by both regressing towards the involutions of a childhood or of a life they miss but will quickly leave; just as efflorescence or effervescence thus designate processes marked by this desinence said to be 'inchoative', an adjective that designates a beginning, here of flowering, bubbling or emotions; just as an arborescent plant little by little takes on the ramified form, bearing or appearance of a tree ..., so a process of hominescence has just taken place by our own doing but doesn't yet know what humanity it is going to produce, magnify or murder.
But have we ever known this?

I therefore call this stage of hominization hominescence to mark its importance and nevertheless play it down in comparison to other, more decisive, great moments; this word rings like a kind of hominization differential. In order to think this differential, I am attempting to dig underneath the time of history towards the times opened up by biology and the exact sciences. Our social knowledge would lose out by cutting all ties with them. In fact, we don't merely communicate in cities and amid concerns that are of an economic, political or cultural order, but our bodies also live in the world in company with other species and things. Already immersed in several types of space by our communication networks, we likewise plunge into several times, certain ones of which are counted in millennia or even, when evolution is at issue, in millions of years. In order to understand a few current events, let's put them back into the long periods assessed poorly by our fathers. In proposing this word 'hominescence', I am attempting to apprehend the newnesses assailing us today in this immemorial light.

Our world and we are going through a crisis spanning millennia; we are suffering from the pains of a childbirth without equivalent in what we have been calling history for not too long; we may be giving birth to another humanity. Nothing can arouse our concern more than this arrival. The least of our thoughts, the most humble of our actions are drawing today, little by little, the silhouette of this particular humanity and deciding in real time how future generations will survive. When bioethics, for example, asks itself 'what is human' or 'what acts are human' without knowing how to give an answer, does it see that humanity is built by the very acts that biology, among others, never ceases to do, day after day, in a risky way, heroic, exciting and tragic at the same time? The human is not the reference; we build it in time through our acts and our thoughts, both collective and individual; quitting its old status as a metaphor, autohominization enters into practice. Thus the road in front of us doesn't resemble any of the ones History has followed, so it can hardly serve as a support for us: hence this book, which plunges underneath its time, so as to sometimes return there. The word 'hominescence' says these hopes mixed with worries, these emergences, fears and tremblings.

Another anxiety without any simple solution: this same event digs a gap between those who are rich in money, body, food, life expectancy, habitat, free democracy and science, having entered, not too long ago, into the 'immortality' from just now, and the mortals – read this name with its density of hope since only death gives meaning – deprived of all these goods to the point of permanent suffering by a symmetry without balance and in part through the fault of the new false gods. The moment of hominescence compels us to resolve this global problem under the risk of total war, and therefore under the risk of a death that would then be fully universal.

The intuition recently produced by such a bouquet of bifurcations and the fact that this bouquet demanded, urgently, a reconstruction of our cultures and our philosophies has accompanied my life and illuminates this book.

Agen, July 1957,
on the Khabarovsk-Arkhangelsk road, 29 October 2000.

PART ONE

THE BODY

The Argument
How Our Body Changed
The First Loop of Hominescence
Three Global Houses
The Greatest Contemporary Discovery
Ego: *Who Signs These Pages?*

1 HOW OUR BODY CHANGED

Unexpected victories

Leaving in the morning for his appointments, the family doctor took in his little bag the effective medicines made available to him by the era before the Second World War: eight to ten, not really more. From the 1950s on, a car would no longer be big enough to transport them. Discovered between 1936 and 1945, sulphonamides and antibiotics, growing in usage, transformed infectious diseases that were often deadly up until then into brief episodes of fever. The two scourges that populated the doctor's offices with cases of syphilis and tuberculosis suddenly declined. Fairly rare up until then, concern for hygiene spread into populations that, accustomed to having horse bedding in town and sleeping not far from cow litters in the country, cared little about cleanliness. The dictates of public health imposed vaccinations and prevention. Later, psychotropic drugs appeared; chemistry was able to regulate procreation and, as they say, liberate sexuality, particularly for women; surgery followed precise medical imagery; we became attentive to children's nourishment ... Some people, like me, remember the time when no one tallied the thousands of dietary poisonings per week, while today merely ten per month will scandalize the media in wealthy countries.

In brief, around the Second World War, medicine began to cure, something it had never really succeeded at from Hippocrates to Galen, Laennec, Jenner and Semmelweis. Suddenly effective, it completely changed our relation to health, suffering, life, death, in short to our body and to ourselves insofar as pharmacy furnished an increasingly open and varied range of suitable remedies, in particular analgesics, antalgics and anaesthetics, which eased pain and sometimes made it disappear. Before the middle of the twentieth century, expert descriptions of diseases and lucid diagnoses outmatched treatments: the practitioner understood the pathologies well and even, thanks to X-rays, saw their lesions better and better but rarely cured. He can cure today to the point that the patient demands, sometimes under threat of lawsuits, his return to health. Imagine the happiness of bodies: formerly rare, now frequent, recovery becomes a right, and disease, once a daily thing, becomes intolerable. In the universe of pain and the incurable, the old-style doctor remained a sorcerer, even a demi-god; as soon as he starts to save people, society – oh, paradox – transforms him into a criminally responsible party.

This unprecedented revolution came, in its entirety, from a threefold alliance between the practitioner (the thousand-year-old adjuvant for individuals), the hundred-year-old scientist (the laboratory discoverer of substances and laws) and the more recent public health services (in charge of statistical prevention), not to mention international institutions such as the Red Cross, the World Health Organization and later Doctors Without Borders. The first one applies to immediate families the recommendations and remedies approved by the second one in international congresses and which the third ones make compulsory by enacting collective regulations. Medicine moves from the individual body to the collective and returns from the latter to the former by the paths of research and public administration. The result: life expectancy regularly increased all the way up to the recent figure, literally unbelievable, of a trimester per year. The population grows older: this verb phrase no longer has the same meaning. Not only does the body of the individual change, but the aspect of society does so as well.

The fact that, in the face of the health poverty of the third world, the probable recurrence of infectious diseases, the diminishing returns of public administration, the power of the pharmaceutical industries ..., melancholy and criticism often win out doesn't stop the fact that this victory report enchanted the years I'm talking about and marked the end of an epoch, whose beginning dates back to the beginnings of humans and even of living things, over the course of which no one knew how to defeat the majority of diseases. Faced with new deaths, the life expectancy of a completely different body increases.

Hygiene and labour

The blessing of hot and running water above the sink joined with the lights of electricity, which the beginning of the century sang of marvellously, transformed our habitat in two ways: less freezing, homes were equipped with toilets and bathrooms. With a twofold shiver of ice and disgust, my body still remembers interiors in which, dressed as for the outside so much did the cold rule there, the delicate residents would wash themselves on the days of major festivals; the others would wait for their marriages. The ceremonial rite of doing the laundry would return with the springtime, for it would take all winter to accumulate the necessary ash; between the icy walls, we would warm our beds with '*moines*' [monks], built with oblong pieces of wood between which we would slip a brazier so as to sleep in nice sheets.[1] The history of medicine also dates the decisive moment when practitioners got out of the habit of systematically sending their patients to the hospital, where comfort and cleanliness prevailed by far over the living conditions at home, to the Second World War; the improvement of these latter conditions caused them to in turn prevail over the living conditions of the public hospitals, in which several iatrogenic and nosocomial diseases were already beginning to appear. This reversal was a historical milestone.

Leaving more welcoming homes not as early, companions got to places of work that were transformed by machines; hard labour saw the number of its forced-labour convicts lower with motor power relieving the farmer and the artisan of the difficulties of lifting, drilling or transporting. The word 'worker' no longer has the same meaning when arms leave off the pickaxe handle so that fingers can push buttons; the peasant quits oxen, plough and yoke to drive a tractor. With cramps in my shoulder girdle, my body still remembers the early risings before dawn and the heavy trucks to be loaded with rocks by shovel or else a nine-tined pitchfork. Now, the body doesn't sweat so much as it pilots or drives. Those people our English-speaking friends call blue-collar workers, having moved into the tertiary sector, became white-collar workers. As early as the end of the nineteenth century, Jules Verne's novels already no longer anticipated changes in forges or mines but imagined machines intended for relations: balloons, planes, telegraph, submarine, television … In brief, Western humanity abruptly moved, during the 1960s, from means or forces of production to communication networks; the beginning of our century made the victory of the global internet and mobile phones complete. Hermes, the god of intermediaries and translators, the Angels bearing messages and their incalculable number, took the place of Prometheus, the old solitary hero of fire. Thus, our body alleviated its pains.

The old body: Height and duration

Today, a little girl of eleven would be too big to fit into the costumes of the French Academy from the early years of the nineteenth century; we can guess the presence and physical strength of Napoleon's *Grand Armée*, which Victor

Hugo's bombast evoked to be like giants. Thanks to the measurements of compulsory military service, we know the average height of French conscripts changed from between 1.55 to 1.60 m during the years 1880–1890 to 1.67 m in 1940 and to around 1.78 m nowadays; embarked, at the end of the eighteenth century, for one of the most atrocious gulags in history, the first Australian convicts were no taller than 1.5 m. Likewise, a woman's first period, which used to appear around the age of fourteen, today starts at twelve. Can we imagine how romantic relationships were shifted in time and age during eras in which the old over-possessive greybeard of *The School for Wives* admitted to forty years and Balzac described *A Woman of Thirty* as finished? In 1833, the Octave of *Caprices de Marianne* says to her, a beautiful young woman of nineteen summers: 'So you still have five to six years to be loved, eight to ten to love yourself and the rest to pray to God.' So Musset reckons about like Balzac.

With life expectancy growing regularly, do we realize that in speaking, for example, of the family and of marriage, we are no longer evoking the same institutions as our predecessors, for whom couples on average lasted six to twelve years, whereas our couples can continue on for more than half a century? I would even wager that the recent explosion in the number of divorces wouldn't stop the married people of today from remaining united, in sum, longer than once and formerly. Likewise inheritance, whose great expectations are so often described in classical literature, no longer has the same reality or the same weight when the heir waits for it less than a decade or more than fifty years; beyond a certain waiting period, the money loses its interest.

What are we to say about the hero who offers his life to his country at around the age of twenty-five, when he has no more than five to seven years left to live? Would we find, at the same age, the same heroism when he may have several decades of life in front of him? The children of the third world who carry, slung across their shoulders, small arms sold by Western industrialists we are surprised haven't been condemned by an international tribunal for crimes against humanity, do they now seem heroic or sacrificed monstrously? When it's no longer a question of the same body or of the same vital time, when the streets, the reunions and the nursing homes are populated with old people, once so rare, do we still think, do we still organize, do we still feel the same emotional, patriotic, legal ..., in short, the same social and cultural phenomena?

The anthropology of pain

Do we suffer in the same way? The greatest monarch in the world in his time, Louis XIV, surrounded by the best doctors in his kingdom, screamed in pain every day. Impoverished and without any help, what then did his subjects have to endure? Conversely, practitioners today sometimes meet with elderly patients who haven't yet suffered from anything. Having in part become responsible for our bodily health, we have more power over our bodies than the most powerful man in the world, disfigured by the rictuses of pain, ever had. In his day, one lost all one's teeth before the age of forty; in the countryside, my childhood often saw and heard these mouths, singing Occitan without dentals. At the end of the nineteenth century, a third of Londoners were suffering from syphilis.

I am mixing dates and facts on purpose to resolve the question: Whose body is it a question of? So I chose the body of the king of yesteryear, in its excellent place, to show it to be impoverished and let the glory of today's common body be seen: this double maximization in number and quality causes invariants beneath the social differences to appear. Hence this, at least statistical, stability: the life expectancy of a woman in the eighteenth century had not significantly changed for centuries. To find more severe numbers, one has to go back to the Neanderthals: 50 per cent stillborn, half of the survivors dead before the age of nine and the rest rarely reaching two decades. Here no doubt is a perenniality implied by philosophers when they spoke of human nature: Weren't they right, unbeknownst to them?

So this is how the daily bodily landscape we all met with before the revolution of the last century was set up and coloured, a crowd painted and drawn by genius witnesses. So don't call Velázquez or Goya, in the past, Daumier, Degas or Toulouse-Lautrec, recently, caricaturists; no, their paintings show what they saw: faces and bodies sculpted to death by suffering, hard labour, hunger, cold and privations, incurable diseases and visible wounds, which the newnesses I am describing have made us forget all about. Stated in broad strokes, that was the era of bodies in the times of pain.

Previous moralities

I attest to having lived, at the end of the 1960s, through the crevasse separating the generations shaped by constant suffering from the one that is suddenly scandalized by a tiny irritation. Consequently, how can we take the recent criticisms regarding the 'doloristic' moralities of our elders seriously? From having forgotten, the very day after they were lifted, these ancient constraints, we commit grave injustices and display a foolish incomprehension when holding forth on the customs, moralities and even the religions of our fathers, forced to give themselves a sustained training in suffering, on pain of not being able to brave its inevitable and daily cruelty. Who can now understand the austere precepts of the Christian or Stoic wisdoms, the high techniques of the body of the medieval mystics or even the Buddhists, all of them confronted with the pain and hunger of each day? The search for ataraxia or contrition depended on fearsome conditions whose insistency has been forgotten by the new body. Our contemporary health therefore liberated itself from this harsh fate fairly recently and in such a way that we broke with what our health had been ever since our origins: a cut so decisive that it closes an era whose beginning we don't even know except in myths and legends concerning the first gods. A contemporary of this abrupt change, I remember the customs in use in the other epoch perfectly and still follow them from time to time, in particular in relation to pain, and I laugh to hear the culture semi-necessarily brought about by it condemned by my contemporaries.

The emergence of the new denuded body

Thus, the new work conditions straightened up the back; the hygiene of home life and a diet that's kept better watch over smoothed the skin; heating undressed us, and we dared to exhibit a body less ugly from the traces of sufferings and diseases. Intended in the past to veil a few visible imperfections, the way people in the Renaissance concealed the Venus collar unleashed by syphilis with the ruff, clothing fashion suddenly consisted in unveiling she or he who no longer had any shame in hiding nothing. For the first time in its history, Western humanity could be seen naked on the beaches. Venus, it is said, was born in the past from the waters, and Botticelli depicts her, being reborn over the waves, dressed in a heavy braid revealing her nudity. Yet, in the same pleated waves, in August of the present years, rose the new body of men and women. The pantheon exploded in number, and the myth became incarnated. Of course, the so-called and sudden consumer society immediately produced obese people, but on the other hand, more tolerant, it no longer hides the bodies of its handicapped, which the shame of old concealed.

From Phidias to Houdon, the plastic artists showed us in glory huntress Diana or muscular Hercules; we see them today, men and women, quite simply, in the stadiums. Abandoning the ideal, a certain beauty becomes incarnated. Yes, for the first time in history, quasi-divine bodies run, jump, wrestle and play before us and, according to the calculations of Alphonse

Juilland, advance in records at quasi-predictable dates. The birth of sports and their popular success largely come from the collective and global rites engendered by the progressive emergence of this body, whose performance grows because it has just been born and whose wrestling matches may replace wars.

Of course, these changes, although important, remain slight in comparison with the strong stages of hominization: the loss of heavy body hair, erect posture, the discovery of fire, the invention of the first tools … This is why I chose the differential term hominescence. In addition, although invisible, the lengthening of life expectancy or the statistical alleviation of pain contributes to a different apprehension of time, projects, life and the world; how can it happen that the ageing of the population appears to eyes exclusively trained in economics to be the weakening of a group, whereas it stimulates education, culture and the coming of a wisdom, which a perspective that's merely economic forgets to the point of only retaining of human life what isn't worth being lived?

Restriction

I'm not saying that we abandoned all constraint during this moment of hominescence. On the contrary, the third and fourth worlds are perhaps suffering even more because of our actions than we ever suffered. Of course, pharmacy, still imperfect, doesn't bring the same benefits to everyone; of course, the unjust violence of money still prowls around us, searching for someone to devour; of course, the last century lived through abominations before which our entire history trembles; of course, medicine, criticized, even vilified, in short, once more at a crossroads, pays a price for its victories, in microbes that have become resistant again, in its residual ignorance, in the financial power of pharmaceutical industries and in drug mafias, in the administration of hospitals, as ponderous and imbecilic as administration is everywhere else. Of course, murder still dashes about, abominable, perhaps invariant in every collective, always as difficult to master, today celebrated every day as a spectacle; in short, we haven't won, and far from it, the entire match. Only a singular naiveté could contest the heavy constancy of the problem of Evil.

The fact nevertheless remains that the revolution took place, that it drastically changed our body and the relation we maintain with it, a habitat so new that we have lost all memory of the one possessed by our ancestors, even our immediate ancestors, and even more so our distant ones, as well as the customs which, by constraining them, adapted them. A number of our new defeats can even often be explained by the recent victories: does the worrying demographic explosion, which also enters into the great events of hominescence, not come in part from the happy lowering of infant mortality? The criticisms of medicine, so frequent today, whose number and repetition darken opinion under their veil of melancholy, also come from them: I repeat, the wealthy are complaining about being comfortable.

Hence the change of ethics. Our old moralities trained the will to live within the inevitable constraints of suffering and early death; the new morality emanates from the freedom acquired against them. We are in part becoming responsible for the duration of our lives and for their quality. Maurice Tubiana interprets the sudden increase in the number of drug addicts (1965: 3,000; 1975: 150,000, in France) by means of the lifting of these constraints, which in fact occurred on these dates. Certain cancers depend on tobacco and on alcohol; cardiovascular diseases depend on diet and on exercise; sexually transmissible ailments depend on often intentional behaviour. Philosophy for centuries had difficulty in defining freedom exactly: when a hundred pathologies vary with our decisions, it becomes incarnated in the entire body. Become its own doctor, the body doesn't fight so much to free itself as it chooses or rejects early death and health. This is the thanatotechnological era.

Moralities of immortality?

This new body appears in our icons for fashion or sports, in our demands for health or appearance. When our ancestors were gaping before Hercules's musculature or invoking Aphrodite's high beauty, they were at the same time measuring the insurmountable gulf separating their state, one dug by suffering and famine, from the state of the gods, drinking ambrosia during their banquet of immortality; their daily pain reduced them to the status of mortals; thus, at a distance from their dreams, they gave themselves this name. If this picture has just been inverted, have we laid our hands on the drink of immortality?

Admiring the bodies of the gods, our ancestors endured their own body, which didn't depend on them. Responsible for the health of our own bodies and for their appearance, which we can in part transform via diets, exercise, drugs or excess, we are discovering the span of its plasticity. Medicine's successes and gymnastic sculpturing make us the partial authors of our corpulence.[2] Effective knowledge and practice change moralities and their constant foundation, which has always distinguished between what depends on us and what doesn't depend on us. Knowing and seeing the body underneath all its scars and pockmarks accentuates freedom and responsibility before its ills, of course, but also before death. Quitting formal prescriptions, morality becomes incarnated but above all changes times.

Once again, a certain 'immortality' becomes, no longer the dream, but the most deafening, carnal and rational project of humanity. The most recent joins here the most archaic, the age of the gods. The history of religion, myth and culture often goes hand in hand with – I don't know why, but I have noted this a thousand times – the history of science and technology. For this new thing, which the preceding pages on death sought, in undertones, to talk about – the possible mastery of apoptosis signals, of that cellular suicide whose signals sculpt our body – will refer to immortality again if it accentuates the increase in life expectancy; but this increase refers, as well and again, to the early epic of the Fertile Crescent, in which the hero Gilgamesh, already knowing that life and death were never separate, went in quest, in his travels, of a certain resurrection: death, we are continually pushing back the executory date of your law; death, we are hunting down your victory; we are putting a button on the foil of your sting.

Decisive dates

We continue to construct this new body by dint of knowing it and exploring it through the medical images granted by ultrasonography, dynamic scintigraphy, scanning devices and nuclear magnetic resonance, whose representations search its levels all the way down to dwarf size, from the organs to the tissues, cells and molecules. Long ago and fairly recently, we no more saw it, inside and in its individuality, than we saw our world in its entirety. Its newness appeared to us at the same time as that of the blue planet photographed from inhabited satellites. When one's own body and the entire world appear differently, how could the inhabitant of these two hotels not change?

We have just transformed our two human houses: the universal niche and the singular habitat. This innovation refers to the first moments, difficult to imagine, in which the body of the *sapiens*, leaving the animal *habitus*, became a being-in-the-world. How? Why? Will we know someday? But we do know several repetitions of these initial or legendary instants: first, when two Presocratic physicists drew the first geometrical models of the cosmos and the Hippocratic school taught anatomy; next, when, at Leuven,

Vesalius, cutting up several corpses, made anatomy more precise in 1543, the very year Mercator gave maps of the Earth by cylindrical projection and Copernicus set up the heliocentric sky, an innovation also taken from the Greeks. Admire with what luminous faithfulness our time repeats this sequence: we explore the body down to its molecular foldings, contemplate the bluish and finite roundness of our common terrestrial vehicle, attain, towards the end of the Universe, the Great Attractor, situated billions of light years away, and mark out the residual light from the two seconds following the big bang.

The same gesture characterizes four events, the first one being archaic, not far from our origins, the second one being antique, contemporary with the Homeric poems, and the third being ancient or, better, defining the Renaissance, with the last one being current. These gestures change the habitat of the body and the view of the world, the view of the body and the habitat of the world. More than the advance of specialized knowledges, they cadence the very process of hominization, therefore evolution more than history, because it is a matter of our two houses, local and global, individual and universal. We no longer inhabit in the same way. We've moved house four times. The fluid expanse of time becomes incarnated in our own bodies and in the Universe. We see this expanse percolate as soon as its bifurcations mark our double habitat, inner and outer, carnal and global. So I'm not writing here another history of medicine, of medications or of health; I'm drawing the anthropology of a corporal change that concerns our evolution instead.

Collective bodies

We don't know the individual body merely through scientific or simple images, but also through numbers and collectively. Statistics plane down the variations proper to singular living things. The mathematical elimination of chance allows personal bodies to be known via a detour through the global; nothing could be more effective than these calculations for mastering cures and administering the state of health of a numerous collectivity. Thus, decisions regarding public health often act with more effectiveness on its improvement than the individual feats, remedies, surgeons or biologists celebrated by advertisements.

But, in addition, politics encounters morality there. These general tallies never fail to be unaware of just how much social injustice deprives the poorest of the benefits regarding pain and death, that a gulf always separates mortals in pain from the rare elect launched on the quest for immortality. Again, the history of myth, culture and religion explains the way things stand better than the history of science or even than history full stop. So the distance between the gods and humans, such as the ancient Greeks measured it for example, substitutes its archaism for the modern difference of class; the scandal increases from this. At low latitudes stand mortals, for whom the tradition reserves the noble name of humans; at high latitudes stand immortals, who incessantly tipple the drink ambrosia. This evident fact can't leave so-called democracies in peace; in adorning themselves with such lying publicity, which no one believes, can they still sing the praises of the most savage and inegalitarian of aristocracies, the most implacable of them all since corporal? The hypocritical discourses make us choke with indignation as soon as their visible contrary is displayed: faced with each obese person of plenitude, the skeletons of the third world cry for blood. Tomorrow, an inexpiable war, Darwin style, will bring these billions of emaciated bodies into conflict with the millions of dollars whose gifts are added to the millions that are stuffing people who are already fat, proud of their exclusive knowledge acquired by means of this pile more than by talent. Leaving ideology, politics becomes incarnated.

The narrative of an example

An example: around the 1970s, the World Health Organization (WHO) planned to eradicate small pox from the world and succeeded in this. For thirty years, we have no longer seen, anywhere, people dying with their skin devoured by suppurating pustules. Literary, historical or philosophical, maybe even mythical, few stories attain the intensity of this exemplary adventure that started with Edward Jenner, an English country doctor who invented the vaccine towards the end of the eighteenth century and ended with the total disappearance of this deadly disease from the Earth. So a few men and women of good will, whose names occupy our memory, spread into the space of the nations and vaccinated every inhabitant around the epidemic sites. With a few years of effort, they destroyed an entire species, that of the corresponding virus. This event resembles the extinction of the dinosaurs and is a counterpart of the fabrication of thermonuclear bombs. Whereas ethics prescribes rather forgetting misdeeds so as to retain the memory of good deeds, where does our passion for reversing this precept come from? We celebrate the victors of wars and the detailed actions of killers but remain silent about this narrative from the 1970s.

Of course, I'm not unaware that this victory, temporary, can soon give way to an even more grandiose defeat; a recent report from the same WHO regarding infectious diseases worries in the face of a resurgence of ailments caused by single-celled organisms resistant to medicines, medicines we are not able to renew as fast our dwarf enemies are able to renew their cuirasses. They are growing; our shield is weakening; the struggle continues; we may have only experienced one of its episodes. We don't easily eradicate 3-billion-year-old bacteria or viruses: they have seen worse on this old planet!

Beginnings of *Homo universalis*

Even if temporary, this victory nonetheless attests to an arrival, that of *Homo universalis* and of its power over the world. With hammers, sickles and drums, *Homo faber* till recently exercised its effectiveness here and now; hitting a nail, following a furrow, resounding in the town square, the action of its tools was limited to a place, dying out at the end of the vibrations, of the season or of its wearing out. Here, by collective decision and reasoned programme, *Homo faber*'s successor extends a medical gesture of prevention to the ends of the planet (this is for space), vaccinates the totality of humans concerned (this is for the race) and causes a species to vanish (this is for life and time, for evolution, in sum). Fulfilling the global conditions of the inert, living things and human existence, its action becomes universal. This is how this narrative uncouples from history and suddenly talks about anthropology and evolution.

Because it tells one of the origins, a contemporary one, of such total actions, I readily consider the narrative of the eradication of small pox to be mythic. We can no longer be unaware that there are or that there have been emergences since we have lived through a few, this being one of them. Are we forgetting the great narratives of the tradition because, contemporary to true beginnings, we are living, without suspecting it, through new myths, not in the recent sense of mystification but in the more profound one of authentic origins? For the adventurous narrative of the WHO doctors mixes the social legendary – since it is a question of radical emergences – with objective science: in archaic or theoretical language, it unites *muthos* and *logos*. So the newness consists in this union of science and narrative, of knowledge and pity. This mixture of objective knowledge and collective epic shows that humanism can become universal. From which a second strange marvel comes: *Homo terminator*, in this case, applies its destructive capacity to healing those near and far. Here is, again, an origin that is the counterpart of the origin designated by the atomic bomb. Since, lastly,

Homo universalis is seeing its responsibility for its death and health grow, as I said earlier, it answers more and more for itself, but also for the world.

Small pox viruses still remain

The collective of the same heroic narrative had promised to destroy forever the last viruses collected beside the final patients. It even set the date for the definitive purge. But even before the airtight bolts of the sterilizers in which the extinction was supposed to be carried out closed, a few biochemists had noticed, in sequencing the DNA of this viral variety, that certain of its genes were in possession of keys to the human immune system, as though, in the course of evolution, they had stolen from us the secrets that made them so dangerous. Consequently, by substituting their own elements for the patient's defence mechanism, by making the host organism believe, through this mimetism, that they belonged to it, they behaved like perfect parasites. Those who eat your flesh gain a decisive advantage if they persuade you they are helping you; mafias swindle and kill those they terrorize while claiming to defend them; thus states amount to legalized mafias. I have said and will say even better to what extent I have promoted this parasitical operation to the dignity of one of the secrets of life, at every level, public or microscopic … In brief, an animal so precious therefore couldn't die without harm to ourselves; so biologists preserved a number of them, as a benefit, in two or three safes scattered across the planet.

Parasite, symbiosis and natural contract

So observe that starting from one of the most ferocious viral parasites in its history, *Homo universalis* brought off here, in a first that hasn't yet been observed as such, an original symbiosis, original because external and technological, between its immune system and the ancient enemy that has become, by means of the sciences, a productive partner. Up till then, we had survived as the heirs of ancestors who had withstood past epidemics from having developed in their organisms a few effective defences against the infectious agent, therefore from having substituted an intra-organic symbiosis for their parasitism; the others died. We still adopt this practice, except that it takes place outside of our bodies, except that we replace the unknowns of life with objective knowledge and experimentation, except that we substitute the definitive healing of the human race for the multiple deaths of natural selection during epidemics. The new symbiont now haunts laboratory work benches.

What are we to call this new symbiosis if not exo-Darwinian? Suspending the total eradication of the virus resembles a peace treaty, at least an armistice, in sum resembles a *Natural Contract*. For we sign them the moment we let it live and consider its DNA as a source of gifts if not of data, as more precious than dangerous, better, endowed with all the more value for entailing risk. We keep it and let it survive so that it can ensure, through our research, a part of our survival.

And to finish, let's dream a bit: once and in the past, did this virus itself let a few humans survive, our fathers and mothers, in order to keep in its possession a stock of food and information? I share with Lynn Margulis a secret admiration for the strange intelligence of these animals without brains. An incredible symmetry: and what if, in letting these individuals survive in safes, we were imitating their ancient behaviour? I think about the *Natural Contract* again: I'm partial to the fact that every kingdom of living thing writes its genes in code, a word signifying both numbers or letters and the regulations of law. When we are able to decipher the real name of the living and things, as well as our own name, we will all be able to sign this Contract in reality.

The liberation of a slave

But let's return to the anthropological or evolutionary body. From the origin and whatever the culture it may have participated in, it had to constantly endure needs that were most often unsatisfied, sufferings rarely calmed, diseases never cured, ten stiffnesses and heavinesses made flexible or lightened with difficulty, as many constraints nothing could be done about, finished off by the inevitable death that breaks youth prematurely. Likewise, the young of animals disappear in large numbers, devoured. This is how we must understand the universal respect for the old, highly rare successes, body miracles standing up to a thousand ambushes and temporary victors over an adverse fate, natural memories therefore of know-how and cultural tradition. Shot through with these weights, these lacks, with all the foreignness of pain and of desire, our body incessantly experienced an essential alienation. Delivered up to drought and plague, never did it belong to itself. The strongest among us stole, in addition, the bodies of others, bodies given over to hunger and disease, as though human ferocity crowned the other ills. As a result, the body was given to kings and to groups. An incarnated slave of nature and of its culture, the body lived like a serf or the colonized, the first servant, the immediate and primitive slave tyrannized by the mind, the soul, the will, the tradition, power – so many important things. Recent philosophies and older religions in fact draw this profile of it, whose pathos comes less, as is believed, from those who spoke about it than from the real conditions of life. Everywhere and always, it met with the impossible. The wisest prehistorians even claimed that its disappearance was always less of a problem than its survival. How, under these conditions, can its hopes not be pinned on a different agency?[3] The body has suffered so much it indeed merited a soul.

So the recent and rapid transformations tend to liberate this multi-millennial serf, whose slavery seemed its quasi-natural state, who had to learn an unforeseeable reappropriation in a few decades. It would be surprising if multiple tails of the ancient servitude didn't remain in it. Lifting the shackles of these impossibilities breaks the egg from which it emerges, stretches and shakes itself, like a bird that has just been born, like Aphrodite standing in her marine conch shell. It metamorphoses and tests its limitless capacities in sports and aesthetics, desires and journeys, food and reproduction, medicine, biological science, genome technologies … Delivered from hunger, it also stuffs itself and becomes obese. The svelte gymnast and the bloated Michelin Man result from this same lifting.

It moves from the impossible, cut up by a hundred necessary constraints, to the open spectrum of every possibility. It inhabits this potential and this new contingency. It is virtual, according to a definition with no end [*fin*] or border, therefore contradictory in appearance but entirely given over to potentiality and capacities. Everything is beginning. When medicine, pharmacy, health policies, the technological adjuvants of work … progress in the dismantling of its impossibilities, we can finally pose the question: What is the body? Answer: it is not; it was, but it is not any longer; for it now lives in the mode of the possible; only a modal logic allows us to apprehend it; it leaves necessity to enter into the possible. This is the best definition that can be given of it: an incarnated virtual. Was theology planning to express this newness when it discovered the image of God in it?

The loop of knowledge

When Descartes wrote that the soul could be known more easily than the body, we didn't yet understand, in reading him, that his philosophy was unwittingly dating an epoch in the process of hominization, for he described there the state of a ship whose hull, still encumbered with limpets, shells and parasites, remained opaque to clarity; blind, his medicine and physiology talked about a strange and poorly known machine because, precisely, a thousand constraints, including feelings, imprisoned it and enveloped it with obscurities. We only approach the body by undoing the multiple cuirasses of suffering and dividing walls of privations that it endured over its long evolutionary course, punctuated with famines and deaths. It was even necessary, for such approaches, to descend into the Underworld (since the body appeared in the dialogues of Plato – and rightly so, for these particular reasons – as a tomb) by traversing the marble of its funerary slabs. These dark defences, determined by the difficulty of living, made the old body into a mummy wrapped tight in its bandages and the ten walls of its mausoleums. This is why we have lived the recent years as a quasi-miraculous process of resurrection. It no longer has any need for linens or spices.

Stemming principally from the sciences, a hundred pieces of knowledge of all types therefore contributed towards undoing a number of constraints weighing on it, pains, privations, diseases, hard labour, and preventing us from knowing it in a freer functioning. New mentalities, an adapted economy, and social behaviours ensued laterally, in short new niches for these new bodies, niches that in their turn and in return had an effect upon these latter. So, knowing diseases and treating them conditions knowledge of the body but doesn't exhaust it since the body consequently changes along with its environment. The newness I am describing consists in finally producing a relatively healthy and translucent body; starting from this newness, everything changes.

So thanks to this relative transparency, recently acquired, we can, although no doubt partly, recognize the body itself to be the major instrument in the acquisition of knowledge. Clearer, it filters clarity. More mouldable, it stocks a thousand software programs with positions, movements, intentions, imitations and adaptations, codings and decisions. It did so, but fettered; it did so, but does it better; it did so, but we didn't see it well. Transparent, it receives and understands; malleable, it forgets and retains; open, it transmits; solid, it knows; active, it decides. This is the new partner in the old problem of the origin of knowledge. But even better: unexpected, it invents. Not everything happens in the brain, far from it. The cognitive sciences become incarnated.

Exercise and training

Fearing lest physical education and sports were being assimilated, certain instructors of these disciplines took umbrage at the fact that in my last book *Variations on the Body*, in which, precisely, I attributed these cognitive performances to it at least as much as to the mind, I wrote: 'Nothing can withstand training [*entraînement*].' But the French language does not reserve this latter term for either one of those two corporal exercises but on the contrary extends it to all human activity. This is how my citation is to be understood. Thus, some pianist confesses: 'If I miss my scales for one day, my playing will become weaker, and I will notice it; if I interrupt them for three days, a few wrong notes will irritate expert ears; for more than a week, the public will desert my concerts.' In saying this, he notes the importance of daily training. How does an algebraist live? He studies groups ten hours a day; if not, he would quickly lose all right to the very title of mathematician. If a writer doesn't write every morning, his hand will grow sluggish. So, it is for every profession demanding an expertise of manual, intellectual, corporal or linguistic execution. Stop speaking Italian, Spanish or German for two years, and you will forget their syntax. The muscles grow rusty, they say, but also manners, vocabulary, style and fine touch.

This truth of the global body plunges into its smallest recesses: if you hide light from an eye from birth, the visual cells and optical nerves will disappear through apoptosis within three months: they will kill themselves through lack of training. Stop thinking, and the neurons concerned will crumble: a daily page of difficult reading rejuvenates even more than morning exercise. The termination of work throws the pensioner into the risk of senility: nothing could be more dangerous than rest. The organism's signals mercilessly put the elements of sensation and of motricity to death when they are deprived of exercise. Thus, the metabolism itself is only properly perpetuated by training.

Raised in the southwest of France, where work doesn't rule as a tyrannical master, to say the least, I have to admit, although with regret, that at the bottom of laziness and inactivity sleeps death. The secret of the work, of course, but of youth and health, nay, of life itself, resides in a steady schedule. Rest, sleep well, but don't slumber too much, or you will grow senile.

Paradox: Repetition produces newness

So here is the necessary condition: this condition on the one hand has to do with the use of language and on the other with the functioning of physiology. But my assertion greatly exceeds this generality of use and function. 'Nothing can withstand training,' I wrote. A mystery of the body, here in fact are its sequences: I can't; I train; I end up being able to.[4] I don't know; I train; I know. I don't grasp; I train; I understand. Up to here, nothing new. But these various series end with another, stranger, one: I don't know any solution to this problem; I train; then, sometimes, invention comes to me. I just said it in a word: unexpected, the body invents. Something like a creation appears. 'How did you find the law of universal gravitation?', Newton was asked. 'By always thinking about it,' he readily answered. No discoverer ever says anything different. I continually place my body before the unknown; suddenly it shows itself; it lets itself be known. Therefore training invents.

It smashes obstacles and leaps over frogs, records, mountains and questions without answers. The mystery that arouses joy and enthusiasm appears to me here: not in the perpetuating preservation of life, of its flexibility and of its functions but in the appearance of something unpredictable, in an abrupt bifurcation, here local but global in all my book.

How does it happen that an act limited to repetition can suddenly transform, one fine morning, the environment, the body itself or the brain, the general state of how things are? When I dedicated my book on the body to instructors of physical education, to trainers and to mountain guides, who I strongly affirmed taught me to think, something which scandalized several intellectuals, I didn't write that sentence without previously thinking about it.

For thought without invention doesn't count; it copies and repeats. It will only find through this training, which therefore is a must for the researcher as the universal model for the conditions of discovery. Inspiration never comes without perspiration. But, once again, training forms a paradox since by repeating the same action, the same research, the same concern that wakes us early in the morning, it makes the body or the world change and promotes the new. No doubt what the professionals call 'being in the zone' marks the attainment of this strangeness. Different expert strategists and technicians of the body, the great mystics already knew, I believe, these secrets; their regular prayers and their mortifications caused them to climb towards Carmel. How are we to resolve such a paradox?

Deviations from equilibrium: The second wind

Perhaps by watching two different rhythms being drawn, the one, regular, that of the electrocardiogram and another one, which seems stochastic, on the electroencephalogram. So in our body stable states neighbour unpredictable series of ruptures and hesitations; does it carry, like the world and the partition of my office, an accurate clock for time and a barometer that's changeable like the weather? Does it live two times then, one of which incessantly bifurcates? The time of repetition, the time of invention; the time, circular, of the planets, the time of contingency?

While Claude Bernard was right to emphasize the regularity of the internal environment, today we stress instead the differences that make it incline. Thermodynamic equilibrium is identical to death; life therefore leaps far away from it. Life deviates from it like the arrows that at random tear apart the waves issuing from the brain. On the other hand, one of the subtlest critiques that can be opposed to classical Darwinism consists in saying that only inadaptation allows evolving, the, no doubt, perfectly adapted species not having changed since the Cambrian.

Like the metabolism, evolution itself therefore combines two contradictory states: deviations and stabilities, beating two times, like the heart and the head, a watch and a barometer. Of course, our bodies adapt to the environment, under pain of death, but they try rather to create a new one; as life progresses in the species, these species become independent of their environment. In the short and long term, our bodies and these species themselves continuously weave horizontal balances and inclinations. This is the reason why the time of the living thing simultaneously follows the regular clock of the elliptic rhythm of the planets, the irreversible and negative course of increasing entropy, but also a strange progress in complexity.

So favouring the search for equilibrium, the repetition of actions, of exercises or even of thoughts would launch, through a kind of compensation, deviations 'entraining' towards an unexpected movement of new adaptation to unadapted actions, then setting up, through repetition itself, a new equilibrium, unpredictable up till then.[5] Even though it designates an almost neurotic repetition, training pulls or 'entrains' 'towards' something other than this redundancy. Through two equivalent terms, the one vernacular, the other scientific, training repeats attraction. So the word would rigorously designate the thing. By thinking about it every day, Newton was therefore subject to an attraction towards the law that the planets entrain each other, by some magnificent tautology![6]

Training causes one to experience the secret of life ...

But, thinking about it in my turn, we never train ourselves in anything but actions or thoughts outside our usual automatisms. To make our stiffnesses more flexible, a gymnastic action often contravenes our habits and breaks them; we repeat it to free ourselves, by breaking it, from this often pathogenic cuirass. Hitting a golf ball twists the body as does serving in tennis. The intuition required by some given strange number or space tears us away from received ideas; running fast and long installs wind, muscles and circulation in a new equilibrium that's far removed from ordinary stabilities. So, in general, training seeks this equilibrium that's outside of habitual stability: so the paradox comes from vanishing so as to let appear, in and through this very experience, the definition of life, whose time inclines out of thermodynamic balance and goes to seek its fortune in this deviation. Life sur-vives, the way the second wind takes over the reins from the first one.[7] So this second life repeats the first one, which would quickly return to death if it didn't go and seek a survival, if it didn't knot together anew the heart's regular rhythm and the brain's arrhythmic leaps: the one pulls the other out of its circle, and the other leads the first one to reinstall itself. Let's leave for elsewhere, this life says, but let's construct, once having arrived, a new residence.

Training opens us up not only to the secret of life, but also to the secret of this book.

... and the secret of hominization and of culture

No doubt life was born from such a deviation, above the inert, imprisoned for its part beneath the bolts of thermodynamic equilibrium. Gymnastics, sports, research work, in short, training, repeat this literally fundamental act. They cause us to survive or to revive in that they draw on the very source of the physico-chemical processes from which life one day emerged in a first deviation from equilibrium. *Existence* exactly designates this deviation and this act.

Bacteria, mushrooms, algae, plants and animals benefit, like us, from this vital dawn which caused them to leave, like us, the stabilities of the inert. How did it happen that they never penetrated, like us, the fundamental secret so as to repeat the same act and leave, once again, from this first shell, then from the second one and so on? So we find them endlessly subject to a programmed equilibrium from which they cannot extricate themselves, whether by catastrophic jolts or gradually, except by the mutations and selections predicted by the theory of evolution. We free ourselves, for our part, from this second prison from having understood, first through the body, how to imitate this first gesture and from having learned to do it again in order to live in a new house, one built on a new equilibrium in deviation from the old one: this second deviation is brought off by training. It lets us leap over millions of years.

Consequently, once begun like this, a series will no longer stop. Hominization consists in this contingent sequence of new deviations, of different equilibriums and new habitats. We have recently moved out of old houses, body and world. So we launch out precariously leaning structures and, if they don't collapse, we build vertiginous houses there. It's always a matter of the same gesture, even if, with each repetition, it becomes unrecognizable. The living had taken over the reins from the inert; survival [*survie*] rises to take the reins from life [*vie*] and so on. We don't recognize in life the exit from the inert, nor in training the freeing from life; nor in invention the repetition of training; we never recognize the new, but it reveals itself over old foundations. We quit evolution; we enter, then, into history. Unrecognizable sometimes, evolution re-entrains history, as today, into what I call hominescence.

Hence the solution to an old problem, the problem of the accord or synthesis between nature and culture. Nothing could be more 'natural' than the gesture of setting up an equilibrium far from a former stability since the word 'nature' precisely signifies a birth and since the process in question describes the birth of life itself starting from the inert, which is held stuck in the second principle of thermodynamics; but repeating the process launches history, the very history separating us from vital, bacterial, vegetal or animal evolution. Culture begins with nature; it is nature itself pursued by other means and become, with each taking over of the reins, unrecognizable. We would never have become the humans we are without training. It opens up to the secret of culture – and to the deciphering of this book.

If you definitely don't like the word 'training', I can substitute the word 'exercise', whose meaning, a related one, additionally benefits from a more than perfect etymology for my purpose since the Latin verb *arcere* signifies to keep away or to deviate, and then, united with the prefix *ex*, to not leave at rest. So I can practice [*m'exercer*] the same demonstration of the deviation or keeping away from equilibrium and thus entrain your adherence. Lastly, what is *existence* if not this deviation from equilibrium or from rest defining in total this disquieted inchoative I call hominescence?

Balance sheet

In short, during the recent decades, a new body was born here. More than historical, this split has to do with anthropology, the evolution of the hominin, the global process of hominization. Of course, I didn't rewrite twenty documented narratives nor pass in detailed review medical care, pharmacy, vaccination policies, prevention policies, health policies, lightened work, the cleaning up of habitats, the expansion of hygiene ... because these fascinating histories all converge towards a result that's more wide-ranging than their components: this anthropological transformation of bodies. Of course, this change benefitted from all the means discovered and applied by knowledge, institutions and humans, whose annals good specialists have written twenty times over, but the result so exceeded said means that while these means can be recounted in historical time, the result enters into evolution.

For the new body recomposes aesthetics, morality and politics, violence and cognition, and even more, being-in-the-world. This rupture took place around the 1970s in the silence of philosophy, whereas in May 1968 the students were chanting on the university campuses of every country a split whose vital importance only strikes the observer today. Who has understood what happened during those days of celebration? Everyone took an event to be political which had quite simply just occurred in developmental biology: the generation of the new body was reaching adulthood. Had a revolution without economics or politics ever been seen?

So I wrote *The Five Senses*, long ago, and *Variations*, not so long ago, not only to celebrate this birth or this advent, but to mark the changes they led to, above all to understand a body that has recently become translucent and visible, finally divested of the cuirass of alienation that had in the past imprisoned it. Wholly new, it demands new things: health, security, long life, food without risk, control of its reproduction ..., in sum, the possible, that very possible it was without knowing it or effecting it and which it has just become, leaving the discipline of being and entering into the supple logic of modes. As, by chance, I have lived in the old one and moved into the new one, bicorporal and bicultural, I know both habitats; since, today, the third and the fourth worlds resemble the house I've left, I know how to speak both languages; I feel, experience and think both ways; therefore I can erect the necessary bridge between these two eras that have been separated by a hard stroke of hominescence. Without any doubt, traversing the rift moves from the impossible to the possible and from the necessary to the contingent; crossing the river mode, the bridge opens on to another freedom.

Fearful contemporaries lament the fact that we are tumbling into a worse alienation by depending more and more, they say, on manufactured technologies: thus the subject, dependent on objects, loses its humanity. Can I console these inconsolables, who seem to know humanity through essential intuition, even though we progress in it contingently? Of course, evolution moves by mutations, but also by selective pressures set up in and through the environment. Begun millions of years ago, the original process of hominization experienced the first of its splits when, in making tools, we began to construct our own environment; certain animals, already, lived more and more independently of their own. When this human environment took on a density compact enough to become a world unto itself alone, it acted upon, as though in return, the very population that had produced it, through a feedback loop that's now well understood. So in sum, we construct our body through the intermediary of the products of our body since technological objects set sail from it. Thus, hominization doesn't resemble so much vital evolution as a production by us; if the word didn't resonate so badly, I would gladly write that it was a matter here of a process of autohominization. We construct ourselves.

Causa sui?

The invention of the atomic bomb, one of the first world-objects, led to the question of knowing whether the human species would survive. What can we hope for? The minimal answer: that this we abide. Some people even ask: In the name of what principle should we work for this specific survival? To this global object, leading to this global question, a global answer inevitably corresponds, one such that it causes us to understand the internal insufficiency of every immanence; for by what right and why would this we, an immanent totality, abide? Without any answer to this question, we are therefore forced to evoke a transcendence exterior to it. This shows the difficulty we have in using global concepts.

But even before we indulge in metaphysical arguments, science itself has anticipated us. Biotechnologies are gradually decoding the various genomes presiding over the reproduction of living things. No doubt the road leading to the complete sequencing of every species will be long, but doesn't this road already serve to fix a meaning for our time? Much as the thermonuclear bomb could be taken as a final world-object, terrifying for its destructive power, we can consider the set of these algorithms to be an initial world-object, that of creation. Soon we shall hold in our hands the birth of the individual, of his fellow human and of his others, of species as well perhaps and therefore the production of ourselves and of our own race. We will cause ourselves to be born, here and now, more concretely and singularly than the broad and collective loop I just described.

From natured, I mean plunged passive into a nature that signifies the whole of what is born or is going to be born without us, we become naturing architects and active workers of this nature.[8] Formerly, Spinoza designated God as *causa sui* or cause of itself: It produced Itself since no creator could be thought above It. We seize this attribute, formerly divine. So the process I earlier called autohominization becomes simply technological: we become operational causes of our lives. In a half century, we have therefore just forged the *alpha* and the *omega* that are the two world-objects, the end and the beginning, creation and annihilation. Through the double mastery of DNA and the bomb, we find ourselves actively responsible for our birth and for our death. Where will we come from? From ourselves. Where will we go? Towards an end prescribed by ourselves. There is where we come from; here is where we are going. This sudden seizure of the two poles of our destiny, specific as well as individual, changes our status. Remaining humans but becoming works of ourselves, we are no longer the same humans. The sudden stroke of hominescence institutes us as causes of ourselves.

Theotokos: She who engenders god

I shall evoke, to finish, an ancient figuration of this new state. A council from ancient Christianity named the Virgin Mary: mother of God. A woman therefore engenders, as her son, her Creator and therefore becomes the mother of her father. It was a matter, of course, as I shall say, of deconstructing the most conventional blood relations, biological filiation and paternity, but especially of countering nature by erasing, as well, through virginity, what might remain of maternity. This overturning of familial relations engendered a symbolic state so new that the face of History found itself changed by it; hence, the birth of an epoch at the same time as of an infant. But I hadn't understood that it was additionally a matter of giving the cause of itself an anthropological figuration. Mother of her father, a produced woman produces her producer. The rational, incorporeal or virginal successes of the human species today fulfil, for itself, this figuration. The Christian era made good on its promise. We are parents of our own parentage.

2 THE FIRST LOOP OF HOMINESCENCE

The exo-Darwinism of technological objects

It took millions of years for birds to grow wings and feathers; in a few months, we build an aircraft. This gain in time defines technology fairly well. The invention of the first tools caused us to leave evolution so as to enter into culture. Through mutation and selection, species appear when some corporal organ or function newly adapted to the unexpected demands of the environment is born. Reptiles flew when the development of lateral outgrowths became wings. As soon as technology appears, we no longer have any need for that long patience nor for a different bodily form and therefore risk disappearing less. Once the airplane is made, we embark; when making a tool is enough, the body changes little if it uses the tool. This is what I call the exit from evolutionary laws: unloading its body of the obligation to slowly obey them, *Homo sapiens* loads its rapid productions with them. To some required adaptation, a blade, a chipped stone or a projectile respond faster than the transformation, random and endless, of a function. The technology-hare overtakes the evolution-tortoise.

Exo-Darwinism is what I call this original movement of organs towards objects that externalize the means of adaptation. Thus, exiting evolution with the first tools, we entered into a new time, an exo-Darwinian one. So this original duration affected these tools in return. Plunging in their turn into another evolution, they transformed in our stead. So instead of sculpting our bodies, duration fashions these objects through the intermediary of our expert hands and our big brain. Does the latter imitate, script or represent the possible adaptation in order to aid this externalization? We may someday know. In using verbs in the present tense, the preceding description may be misleading; the process it follows took a vastly long time and continues on today.

Setting sail [Appareiller]

This stone serves as a hammer in place of the fist, more fragile but serving as a model, and this lever externalizes the forearm … Thus a sort of *appareillage* [setting sail] took place and always takes place, in every sense that can be given to this word which evokes at the same time the devices [*appareils*] themselves, their like [*pareille*] resemblance to body functions and the putting at a distance of these functions, their externalization, that loss of parts of our body into fabricated objects tossed at random into the world. Our vital functions get lost outside in the inert (certainly) and intelligent (assuredly) things, and this objectivization can improve their performance. The wheel goes faster, and without getting tired, than the portions of the sphere the hips, knees and ankles travel while walking. Strangely trinitarian and wonderfully trismegistic, technological objects therefore mix the inert material realm, stone, bronze, iron or fire, with what is called mind, fertile in ends, skilled in means and productive in results, lastly with the functions of life, nutrition, perception and motricity, bearing, action at a distance and, later, reproduction. Thus tools enter into time, and their evolution reproduces, infinitely more rapidly, the evolution that changed our bodies, which, in return, change less.

I imagine that this exo-Darwinian culture was born at the same time the first stone was chipped and that the whole of these settings-sail developed it. This is what caused us to bifurcate from evolution. The human began, in fact separated itself from the other inhabitants of the world, when this decisive event took place, progressively of course: it exited, at that moment, from what philosophers call nature and naturalists call coevolution. From the first tool on we no longer had the same world as animals did. We were starting the construction of our own house, which would replace the world.

The first chipped flint crystal laid down the first stone of this third dwelling, built between the two fundamental houses, body and world. Then the entire building evolved. And we no longer budged from this habitation whose increasingly wide and virtual walls separated us from the other living things, which, for their part, only inhabited a stiff body and a poor world.

The loop of hominescence: Returning to evolution

This is what defines our history. Why give the advent of writing as its lower bound? Because, narcissistic, we belong to a civilization of paper? Because this convention allows us to look down on, by considering them to be outside of History, cultures, more numerous than ours, whose languages aren't written? Graphocentrism, another racism. No, our time, broader, came about as soon as our cultures evolved more and more rapidly than our nature and in its stead, so much so that this nature little by little became forgotten. Our bodies evolve little in an artificial universe evolving, for its part, differently and more and more rapidly.

So fast even that, suddenly, the technological objects, archaically loaded with the time of evolution and protecting our bodies from changing, affected them through a feedback effect and so strongly that our bodies began to change. Of course, we could have, for a very long time, said that our bodies differed from those of animals and the other living things because they inhabited an environment they built for themselves and which, in return, acted upon them. But these protective walls only affected them superficially, softening them, polishing them, weakening them, giving to certain functions a priority they wouldn't have had in a nature less protected by culture. The city dweller lives in a foetal and nervous state, and even in a different time than the country dweller, more perceptive and muscular for his part, and this rural dweller lives differently from the hunter-gatherer.

But no one 'had yet been able to add a single cubit to his height' nor several decades to his existence; we hadn't even figured out yet how to anaesthetize certain pains, nor how to sprinkle the ovum with sperm at a distance from affection; in short, we hadn't yet transformed our bodies in depth the way evolution did in the past. This very morning, we have taken up this evolutionary time insofar as it sculpted bodies through death and reproduction, with these latter now entering into the house of culture. Fundamental coordinates of living nature, death pangs and birth, Eros and Thanatos, shape living beings; even humanity, even this animal that's left coevolution, received these two constraints from outside its house; its culture was subject to them from nature, so named for this birth reason; humanity has just made them enter into its own dwelling. Studied by science, grasped by technology, death and reproduction become cultural. The forces shaping our bodies now come more from the environment we have built than from the given world, more from our culture than from nature. The hominescence described here, defined as a hominization differential, must above all be understood as an autohominization differential. We are beginning to domesticate death and reproduction, sculptors of flesh, the driving forces of its evolutionary time.

We had quit evolution for the entire duration of *Homo sapiens*; we are continuing its course these days but along a new route, one even more divergent from the old evolution (itself always at work in the other living beings) our first exit had diverged from. A second bifurcation, a new history. Of course, we are still subject to its laws, but from having decided selection for a long time and deciding mutation starting these days, we are changing its speed and direction. With regard to the old and natural evolution, we knew neither selection nor mutation since, contingent, it moved randomly through the necessity of its laws. Can we know them for the new evolution, artificial and cultural, since it depends precisely on our knowledge? Where are we going?

Finality

Of course, the first evolution had never known finality. It couldn't even be conceived except on condition of putting all idea of end [*fin*] between brackets. Nature has no goal or aim. It must be remarked that Aristotle conceived the said final causes with regard to technological objects: this book, I am writing it to try to express some truth as much as I can and am able to see it; this weaving, Penelope expands it to clothe her husband upon his return and undoes it at night so the suitors won't put it on. Yet, while these causes didn't apply to the living kingdoms, in reality without finality, they return to centre stage as soon as the new evolution dives into a biotechnological time, understand by this a time that obeys, of course, the general laws of the living, but which is also commanded, launched and carried out by our culture and our projects with our means and our tools. By modifying genomes, by producing reproduction, by genetically modifying organisms that will react in a different way to the environment, therefore to adaptation, we are starting to invent the new time of a second evolution. Do we know how, will we know how to produce this time? By coding new living beings, we will no doubt be weaving it. But why? With what intention and what aim? Towards where will this time go? Today's revolution consists in the fact that the new living things are in part becoming technological objects. Unlike the first living things, the second ones emerge from a plan. Consequently, why, with what aim are we fabricating these new mixed bodies we are calling biotechnological, thereby admitting that they mix end and the absence of end?

It was easy for Auguste Comte to say that science resolved the question *how*, never the question *why*. That it didn't have to pose the latter question since rationality itself was founded on its exclusion. Only metaphysics asks itself *why*. And science, the positivist said, excludes all metaphysics. Tranquilly seated on this very clear solution, we all thought more or less the way Comte did. But did we see that he was assuming that all science could be reduced to the model of mechanics or physics, the very word of which leaves metaphysics out?

Yet, at the very limits of this reductionism, at the extreme refinement of fabricated objects, suddenly appeared, thanks to them, the statistical lengthening of life expectancy, a relative control of reproduction, a fabricative intervention in genetics, all three of which run along the arrow of time. When we invent another time, we cannot just blind ourselves to its arrow.

At least and explicitly since Aristotle, that is, two and a half thousand years, at most since *Homo habilis*, that is, several million years, the question of final causes has in fact been asked regarding man-made objects: I repeat, this axe, the metallurgist forged it to cut down trees; this vessel, shipbuilding erected its handrails for pitching on the lovely sea. Of course, neither the mushroom nor the wolf came into life with an intention of this type, for no one ever knew for whom or for what the rose and the reseda were born, except for the one in whose eyes God, the eternal craftsman, forged mankind so that he can save its soul. Scientists and philosophers, since Darwin at least and Voltaire at most, have therefore detested final causes, particularly since they had made little use of hammers or hoes. Yet, through biotechnologies even more than through traditional agriculture or livestock farming, the living thing is becoming, in large part, a man-made object. We chipped the stone to hunt; why do we clone GMOs? Is it enough to say: to protect ourselves from diseases? Our practices contradict our philosophies.

Quantification: Maximum and minimum turned around

But our practices contradict themselves as well. Never have we had so many means, but, to our aggrieved shame, never have we had so few projects either. The gap between what we could do and what we in fact do with them characterizes our time of omnipotent impotence. We master the atom in order to set obedient police officers against protests by spoiled children; an ingenious wealth of electronics allows us to exchange moronic messages. Reversing this relation, our fathers crossed the seas on board fragile frigates, and more distant ancestors forced their way across the Bering Strait on foot: few means for giant projects.

Not long ago, Leibniz defined science, the world and divine creation as three works in two words: the minimum (of expenditure) for the maximum (of effect). He gave as an example of this optimization a spherical raindrop, a natural form in which the smallest surface houses the most volume possible. Remember, likewise, your experiments of yesteryear: four or five lab exercises, therefore not much data, rare little points scattered over the graph paper, gave you the idea sometimes to extrapolate an entirely general law. In physics class, you thought in Leibnizian terms: few supports, a big bridge. Today, on the contrary, we are piling up data that sometimes reach 10^9 bits of information so as not to truly be able to process them: mountains that haven't yet given birth to a mouse. Did our civilization have the bright idea to reverse the Leibnizian principle, and did it spend wonders in order to be enriched with or accumulate paper? From too much simplicity, the projects of billionaires remain impoverished: so much intelligence for a bank account!

But it's less a matter of quantity than of time and age. Our means date from this morning and our projects from the day before yesterday; they seem to still remember the ended era when we had few means. We pride ourselves on crossing the seas with the telephone or internet, but our forebears did so by means of a square sail rigged with difficulty. We go faster, that's all, without adapting our targets to our capacities. We don't yet know how to make use of the very scale of our powers. We have the means to transform deserts into forests and gardens; we could feed the hungry of the Earth with our surplus; we could cure their diseases – and we sell them weapons so they can kill each other. We pave their hell with our miraculous abilities. Does this 'we' withstand 'our' intentions?

Like certain newly rich people, lagging behind their recent wealth, who maintain a cost-conscious spirit and greedy behaviour, we are behind in seeing our strengths and haven't adapted them to goals that are in proportion to them. We are lagging with regard to projects. Our fathers listened to the opera on the telephone, thinking this device was uniquely intended to spread the music of Verdi to a few living rooms; thus they were mistaken about the scale or number, about the global volume of a capacity. We see that a practical object is of use, but we need time to understand for whom, for what and up to what scale it can be of use.

From metaphor to the concrete, from Narcissus to Pygmalion

From criticism, an easy thing, let's move on to the response, a difficult thing. For the first time in History and probably evolution, the self-formation of man by man has quit the status of being a metaphor. The sciences and technologies have just taken humanism at its word: Do you want to make man? Here are the means to do it. And now, humanists, show your project, for the moment has come to carry it out. We are going to make be born, through knowledge and will, the children of women and men: we shall soon take human 'nature' at its word since we will make it be born. If you don't answer the project question, you will be accused of being a dreamer. Knowledge and know-how confront philosophy with its boastings and its responsibilities. This is why it flees today into commentary, here, and elsewhere into language games. But let's again drop criticism, that easy thing.

'Do and, by doing, make yourself' [*faire et, en faisant, se faire*]; this is a motto whose outmoded scent of morality we smell and whose slightly grandiloquent gesturing makes us laugh. Do we truly understand that we are starting to fabricate our bodies? That the skeletal thinness of these bodies and the obesity of those bodies result from acts for which we are responsible? That we make ourselves sick at leisure, that we add three months to our lives every year and a small cubit to our heights with every generation, that we govern part of our reproduction, that we sculpt our bodies at least as much as apoptosis? Evolution requires millions of years to change a function we cause to vary in the laboratory. Of course, I'm pushing things to the limit; many opacities don't always depend on us and maybe for a long time. Our lives nonetheless are beginning, in many respects, to depend on us. Tomorrow, will we sculpt our bodies from before birth?

The story of Narcissus, a virtual image and one of fascination, used to be helpful for our metaphors. The more artisanal story of Pygmalion is already better suited for us: this statuary, so in love with his work that he wanted to bring it to life, quits myth to become, quite simply, a true story or an attainable end. 'Now speak!', Michelangelo cried at his seated Moses, holding the Tables of the Law in his hand. We hold in our possession the tables of our Code. Whether we want to or not, humanity is our project. All the women, all the males, all the old people and children, girls and boys. Next or at the same time, the other living things. Make them and make them live: Why and how? What does this mastery have as its aim?

Mastery

The master, it is said, had the right to decide life and death. A shameful redundancy: in fact, he only had the power to kill; he accorded existence when he decided not to commit murder. Even he made living things blindly, but could practice death by intention. His power was therefore expressed more through murder than creation. Hegel reduced the tragicomedy of the master and the slave to a feinting or a dodging before death. Death alone then decided mastery. The mastery of life passed through slaughter.

So every project, war, conquest, dominance or occupation was perpetrated by charnel houses, stinking on the evening of historic victories and, not far from the palace, in the shadow of gibbets. Omnipotence shines forth in the number of gallows [*potences*]: with a single word, the language admits as much. In this hideous past, we hadn't left bestial hunting. Who differentiates Alexander, Caesar, Bonaparte … from Stalin, Pol Pot and Hitler, if he isn't evil-hearted? Illustrious men, acclaimed leaders: killers. Heroes: murderers or victims. Famous and mortuary models. When, led by such butchers towards slaughterhouses of this sort, this history is repeated around us, we start to become outraged at it. We have barely left a deadly history, cadenced by the exploits of repugnant persons (from the verb *pugnare*, to fight), captains of blood streaks, businessmen over rivers of sweat.

As a result, we can no longer repeat the old projects of a disgusting history. Thus, we no longer have any today because the ends we have set for ourselves up to now arouse nausea. Might we be, regarding this point as well, in the process of mutating? Previously rooted in death, does power have as its project to become rerooted in life? We can ease life, prolong it, reproduce it: Have we just acquired the right to decide life and peace?

The latter project then consists in making humans, in making them live, in making them live well, and first and foremost in peace. We fabricated tools, weapons and objects, often paid for with the blood of our fellows; we had never made subjects. We had never made humans. We had never made humanity. Not individually, not collectively, not universally. How are we to reach an agreement about the meaning of this enterprise, which has no equivalent? Everything seems clear in the preceding sentences except the pronoun: we. Everything depends on us, of course, but what does this 'us' signify? If we can build humanity, what can we do about this dark collective, placed between the individual and the universal, both of them clarified, both of them thought lastly and soon within the horizon of our practices? The humanist question then runs, with a single movement, from biology to politics.

Chance and intentions

By chipping the stone, we were substituting the edge of the knife and the sharpness of the flint for our weak canines faced with the fangs of tigers and the tusks of mammoths. By leaving evolution, we were already attempting to eliminate chance. For evolution presupposes this chance, which gave us as many and as few chances as other species, but which tends to get erased by the creation of the technological, cultural and human environment. We increased our odds of survival and so ran fewer and fewer risks thanks to our technologies, those anti-chances.

As soon as this man-made environment, as soon as our third house affects in return the first one, formed by bodies, as soon as the bodies, within, become, in their turn, bio-techno-structures, they receive full force the question of the aim, of the project, of the end, of the intention or of the target. Certainly this rose blooms without any why; certainly the colour of her eyes, an unpredictable smile in comparison with the smiles of her blood relations and her unexpected use of our maternal language surprise me in my granddaughter, conceived blindly, but for what purpose do we intervene in the time of evolution if and when we do intervene in it? This technological action, like every other, contributes towards erasing chance. So why are we eliminating chance? This last question even amounts to a tautology since eliminating chance consists in substituting for it something other than it, that is to say, a purpose. Every project substitutes a purpose for chance. As soon as we pronounce the word 'biotechnology', the why is added to the how. We are living through the end of positivism.

Therefore, without any purpose, we make finalized things. Are we fabricating finality without any end [*fin*]? Kant said of this formula that it defined beauty, in a *Critique* in which, not far from this aesthetics, he analysed the living being. So ought our laboratories to produce works in the sense of the fine arts? Maybe. I only see a single objective project here: What do we want to do with these lives that we are beginning to form with our hands? Can philosophy answer the questions: Where are we going? Where do we want to go? I hear today, in a muffled and deafening way, chance being praised: Do we desire so much to go blindly? What horror therefore lies in knowledge? What worry do we dispel before conceiving such a project? Do we desire so much to remain in blindness?

The indetermination of humanity and of its fate

We don't want to answer these questions. Persuaded that such questions constitute our wealth, we want to leave our aims in indetermination, particularly as what is at issue is the human to be born. We don't want to know what life, what time, what human we want. Strictly speaking, I do want to ask myself the question and answer it voluntarily in and for my personal existence, but we don't want to resolve it in general, which would amount in some way to imposing a definition on hominity, a definition that would cause it horror in the short- or long-term. For the first time, we know why we don't want to know why.

Consequently, for the first time, again, we know how to answer humanism's question: What is man? The animal that refuses to know what it is because its entire fortune precisely consists in not knowing this. For the first time, lastly, a speculative non-knowledge seems to liberate us in comparison to a practical knowledge, which we have always affirmed can and knows how to liberate us. For this meta-knowledge would bring sorrow to us and our children; it would bring us back to the level of brute animals and fixed plants, which, for their part, are something defined. We discover the horror of all ontology. So we leave open the indetermination of the answer. What is this humanity? Answer: a possible in a range of potentialities, potency, yes, omnipotency since it can become anything. What is man? This range itself, this omnipotency.

In this respect, the second Evolution, the one that is beginning today with our means and that strange and new non-will, the one we are building, fabricating and ordering, resembles the first Evolution, which was undergone, given, contingent, at random. Both of them move towards the possible. The one naturally, the other voluntarily. We were living a contingent nature; we will be living a contingent project.

Useless and uncertain ontology

What I have just claimed depends on the way the question 'what is man?' is asked. It uses the verb 'to be', an empty and worthless word, undefined and indefinable, a poor auxiliary, substitutable at will, a washed-out token of our languages. This mercenary doubtless makes such a fortune in the books of philosophy because it signifies everything, anything and nothing. This undefined suits charlatans marvellously. Nothing could be more mendacious or stripped of all meaning than ontology, along with its twin phenomenology, both ugly names; giving oneself such unattractive names doesn't plead in favour of anyone. In fact, ontology frantically repeats questions about Being; let's not be surprised if it doesn't provide any answers to them since questions so formulated make a complete trip through languages and meaning without leaving any exceptions to them or breathable space. It even takes great pride and delight in only spinning around in and focusing on questions. There are other places to mutter incantations. Practise, at leisure, the occupation of rewriting this page by substituting the word 'Being' for every verb and a few substantives, as well as substituting 'being' for a few small-time others: this empty and easy enterprise will earn you great esteem.

No, Being doesn't concern us. Being, perhaps, concerns animals, plants, mushrooms and bacteria, sand and lakes, fire and rock, air and clouds running in the riverbed of the wind, although we may not be able to verify it. We exist neither as beings nor as Beings, but as modes. Our existence floats in the square of modality, where possible, impossible, necessary and contingent put up the four walls of our cultural and natural dwellings – body, technology, language, arts and world, narrow shacks as in the past and vast palaces as of now. From our own dawn, we have built our three houses in this fixed and moving, hard and soft, adventurous and regular square, not in the dismal plain of being.

Under the roof of the impossible supported by a necessary truss, to breathe, we open up the windows of the possible and leave by the doorways of the contingent. Vibrating with time, we move, free, towards a hundred possibles under necessary constraints, live therefore contingent moments, sometimes attempt the impossible, and by running into its obstacles, it sometimes even happens that we miraculously create necessity. Inventive of its duration, the hominin goes absent from the time in which Being and beings are involuted repetitively. Living and thinking in this original flexion, we have forgotten beings as well as Being, eclipsed ones of presence, for millions of years – forgotten not since the Presocratics, but at least since *Homo habilis*, since the first dawn of our hominization when the first technological tool set sail from hominian hands.

What's surprising then about not wanting to be since we even are not. Except as unexpected, adventuring, unpredictable, seeking, disobedient ..., therefore possible and contingent, flexing and therefore knocked to the point of drawing blood against the impossible and the necessary. All it takes is someone wanting to determine me in being for me to immediately flee that definition, that proposition, for me to evade outside that injunction. Man wanders with regard to being; it eludes it and doesn't want it. As soon as man rises, its body inflects [*s'infléchit*]; as soon as it walks, it runs, it dances, it avoids, it invents feinting, never goes where we think it will; as soon as it speaks, its verbs inflect [*fléchit*]; prepositions, declensions, in general flexing [*fléchissement*], sculpt every language in the world. The lightning flash of flexion governs hominization. If not, might as well become animals again, might as well plunge back into coevolution. Humanity detests being. And even now I doubt if, revolving in the vague generality of language, ontology will ever even know how to discourse on rocks, backwaters, crocodiles or sequoias. Being-there or 'here lies' [*ci-gît*], ontology only knows cadavers.¹

New man-made objects stripped of finality

The question remains, left hanging earlier, of man-made objects and their ends. In fact, new tools have just emerged, stranger than the usual ones as far as the question of finality is concerned. For an immense difference separates a lever or an engine, on the one side, and a writing case or a computer, on the other, that is, an instrument that functions at the entropic scale and another which processes information. Recording media that store, transmit or receive information serve no purpose, in fact, while a lever serves to lift weights and an engine to produce movement. Yet these media arose at the beginning of History: a wax tablet, a papyrus, a parchment, a stylus …, I repeat, have no finality in the sense that can be given to this word in the case of the lever or the engine. And contrary to what the history books say, history changed more through these tools without any end than through those that had one.

Objection: but they serve to write or count. Of course, but these verbs, these acts, quite precisely modal, are established in the possible: on or through them, anyone can play, work or write anything, a declaration of love or tax returns, a theorem or a poem, the false and the true, an insult or a caress …, in any way, changing numbers or codings, changing equations or music, languages and styles …, an entire unexpected range of narratives and accounts. Even this piano, this organ, this musical instrument …, definitely this calculator, this computer …, this network, this internet …, all this serves for anything, for any result, and therefore doesn't serve any purpose in the usual sense; or rather it serves for the impossible and the possible, the contingent or the necessary: these man-made objects, too, revolve in the square of modality. This is why I readily say that unlike any other tool, always by definition specialized, that is to say, finalized, a computer can be defined as a universal tool, an expression that, itself, can be taken to be an oxymoron. It allows one, surely, to do anything.

In fact, the current pregnancy of these instruments devoid of finality confirms the entire preceding reasoning. For if life recently became man-made, this event came about in large part from these man-made objects. We no longer conceive the organism by means of simple machines, pulleys and levers, as Descartes or La Mettrie did, nor by means of thermodynamic or electric engines, as at the end of the nineteenth century, but rather as information-processing machines. The representation of life follows our technological capacities. Thus the old mechanism, which represented it by means of conventional tools, was exposed to a comical contradiction since it contested final causes while imposing them through simple or complicated machines. A new mechanism replaces it, of course, but by means of man-made objects without finality. This is how we think the living thing, which is constructed today as without finality. This is how our means unite with our projects.

Loss of finitude

Possible all the way up to a certain omnipotency, little by little becoming causes of ourselves and of certain totalities concerning the becoming of the world, we are losing much of our finitude. Almost all the new questions and answers proceed from such a loss. We are more and more quitting the *Lebenswelt* cut out around our bodies, our old narrow shack. Our local acts and projects, our highly defined achievements are in part falling into disuse as the world widens. Entering the global by little steps, we are quitting our old houses, the weak body, local tools, the limited world; we are losing the comfort of cramped habitats and their narrow doors. Tragic sometimes, finitude nevertheless permits many a philosopher certain beautiful theatrical gestures, heroes of the absurd, who never leave the comfort of their rooms. Now we have to disillusion them. Thanks to said finitude, we could define ourselves; the very words repeat each other. We are in the process of leaving both of them at the same time. The absence of Being, of a defined project, of a definition of man testify to our commencing infinitude.

The prophetic lamentations according to which we are going to lose our souls in biochemical laboratories or in front of computers agree on this high note: we were happy in our little shack! We touched its walls; our head bumped into its beams; we lived squeezed together. Ah! Solidarity in the face of the universal non-sense! What happiness: we couldn't cure infectious diseases, and during years of high winds, famine killed our children; we didn't speak to strangers from the other side of the stream and didn't learn difficult sciences. The falling of the walls, the abrupt advent of the virtual and the possible, the progressive liberation of the body, of cultures and the world, the habitat discovered in the floating square of modality, all this makes the heads spin of those who weren't working for this exit (slow for a long time and yesterday sudden) from finitude. All the sorrow they project onto the world comes to them from the fact that they never knew how to leave their rooms for even an hour. Never has the increase of our means been accompanied by such a chorus of regrets on the part of those who have never worked on these means. The extreme difficulty of freeing oneself from this little egg of finitude – to my knowledge, we still have enough of it – explains and excuses the error. *Homo universalis* is starting to live in the open air of this relative infinitude.

And what if our old violence ensued from the constricted narrowness of our archaic finitude? Full of hatred and resentment, sickly and mediocre, we harboured souls of doves that were ready to kill; acquiring a few more capacities offers us an opportunity for generosity, as though to tigers. With the gain of means, do we lose wickedness?

3 THREE GLOBAL HOUSES

The virtual house of living things

Maybe the genome of a living thing, whether bacterium, alga, mushroom, plant or animal, envelops the time of its development; maybe even it contains, engraved, the hour of its death. In any case, as soon as we cause one of its genes to mutate, we cause a new duration to appear. So each genome contains, virtually, a bouquet of different times, as numerous as its genes and their combinations, as numerous as the mutations that can be carried out in it. This deposit in this single bank already counts up to a large treasure, at least a virtual one. But we don't know, and far from it, every species, about which some people estimate that the unknown ones still total in the millions. Let's suppose that we could, one day, draw up the inventory and classification of them, a bold hypothesis but not a crazy one.

We have as yet only decoded a few genomes: a few bacteria, worms or flies, and humans soon. Let's suppose we have finished writing, probably much later still, the vast book in which every genome, finally decoded, of every species, living and extinct, current or fossil, real and possible, would be unfolded, a mad hypothesis but yet not insane. The conceivable lives and the possible mutations would be virtually embryonated in this well, this stock, this bank of banks. In the face of these gigantic bouquets, whose jets gush forth from this deposit, ordinary history and time, such as we perceive or think them, flatten.

Passing to space

I use the verb 'flatten' intentionally. We live and perceive in a three-dimensional space, plus time. Housed there, we experience the position and solidity of these slabs of comblanchien limestone, the petals of this red rose as well as its scent, in brief, sensible qualities. But these change once we go down to the scale of particles, whose position, merely probable, and solidity, lost, seem paradoxical. At that scale, we would walk through marble without any trouble.

What then is the real? It at least unites general relativity, which gives an account of the world such as we perceive it, and quantum mechanics, whose equations explain the world of atoms, since we live in both of them at the same time. The mathematical theory of superstrings gives us the hope of explaining such a union. The ultimate real this theory attains, the real precisely that unites intuitions and concepts as contradictory as the dense resistance of the table and the emptiness in which particles vibrate, presupposes vibrating multidimensional 'strings', called 'superstrings', which of course we don't perceive but which give an account of the global world. If we project these multiple dimensions onto only three dimensions plus time, we find relativity again along with our ordinary experiences, the marble and the rose.

Thus, long ago, on the walls of his famous cave, Plato projected the real three-dimensional world of the outside onto images that were no more than two-dimensional on the wall, and which the prisoners of the cavern, mesmerized by this movie screen, took, in ordinary life, to be reality. So leaving the cave or these appearances towards the real as such consisted in adding more dimensions to usual experience, one dimension for the prisoners, several for us. So if superstring theory succeeds, it will become evident that we perceive and think in a flattened world, less apparent than squashed, that is to say, projected onto only three or four dimensions. This is our new prison. We free ourselves from it by coming out into the broad place deduced and set up by a brilliant set of algebraic equations, the tutelaries or guardians of the real, which unfurls in many more dimensions.

Returning to time

As with this theory, in which reality appears in equations more than in front of our eyes, life as it goes and our history of ordinary living things depend on a contingent lottery, shuffling the cards of our parents' genomes, a lottery which brought this or that combination into existence. In comparing its small flow with the vast book just evoked in which the totality of possible deciphered genomes would be deposited, the time in which we evolve and which we can grasp in fact becomes squashed into minuscule dimensions:

the lineages of my father and my mother, plus the entirety of the families stemming from the survivors of the Great War, of the Black Death, of the great invasions, of Noah's flood …, plus the descendants of the sheep domesticated by Abel and the fruits cultivated by Cain …, plus the species, unknown, that rustle in the rain forests, not many fates, in sum, even if their number does seem enormous.

It suffices then to project the large book I just hypothesized onto a few lines of its combinations, in the middle or at the bottom of one of its pages, to find this squashed time or the little world of our narrow time, which opens out, conversely, on to the broad place designated by this vast set of every deciphered genome. Even though we are flattened in a narrow prison of space and slide along a thin-threaded time, constricted within these two frameworks of appearance, superstrings and this book-stock free us from them. We can leave our little lives to explain the wide world, the one we finally recognize to be real.

The specificity of life

In thus climbing up to time and space completed in their full and unfurled dimensions, these two authentic acts of thought immediately show the specificity of the living thing in relation to the inert. Yes, the real lies in mathematics, or at least mathematics alone opens up to it. The other paths remain within appearances or particular instances, images or reductions, flattened or squashed. Yet who doesn't see that in this matter algebra and its equations, which lead to superstrings, give us the key to the inert real, while combinatorial analysis and its algorithms, which fill the large book, give us the key to life? Yes, mathematics opens up the totality of the real but on condition of distinguishing two mathematics, the Greek one and the algorithmical one, the first one being proper to things, the second one being specific to living things; the one, declarative, sets forth rules, the second one, procedural, puts forward codes; the one unfurls and attains the universal, the other constructs singularities; the one dictates the necessary and predicts it, the other presupposes the contingent without foreseeing it. For, invading the universal, inert nature has no need to code the local; singular, the living thing only invades tiny portions of the Universe by coded varieties. For the inert occupies spaces and time, without limitations, while the living thing inhabits limited places or pockets, and composes rhythms that slow the flow of mortal time. A philosophical question would consequently consist in asking whether two mathematics really exist or whether the one can be reduced to the other. This question is easily resolved: there is only one mathematics, but its two versions differ greatly.

Directly considering the things themselves – a molecule, a crystal, a membrane, a cellular association … – doesn't succeed in decrypting the secret of the distinction between the inert and the living because all things exist or live flattened by the squashing on the part of the two projections I just defined; the only thing that would succeed here would be broadening towards the global real, the one opened up by mathematics. So Plato was right to place the real there, except that he didn't know algebra and looked down on the thought of algorithms as much as he did the slave from the *Meno*.

The third global house

To the deductive realm of superstrings, of the order of the inert, and to the book-sum of deciphered genomes, the treasure trove of living things, we are adding today, for we could construct it, at least virtually, another broad site, the bank of data banks or of available information. In principle, we know how to lodge this universal encyclopaedia, the open set of the sciences (whether pure, exact or applied), of knowledge (whether technological, cultural or artistic), of information and narratives (whether entertaining or uninteresting, useless or pornographic …), and we are attaining it.

For good measure, superstring theory will also rightly be deposited there, along with the global book of set forth genomes. The three places then form but one place, even broader still. We can now outline, at least virtually, the span and profile of this stock, whose informative power may surpass the number of elements of the Universe. Rabelais coined its name 'encyclopaedia', whose meaning hampers me today because it evokes a cyclical form, and without always being able to achieve it, Leibniz, Diderot or Hegel revered it. This stock is today passing from concept to possible practice thanks to our networks. It suffices to project the broad dimensions of this general set onto the restricted dimensions of the rigorous or the futile to find some dimension of knowledge, existence or song. The universal grand narrative of science is included in it.

Thus the things of the world, the living beings and the dead, the arts and the sciences each discover their foundation, where one unites with all the others, from which the set of the possible emerges, symbolic and real mixed together. So there are still a few questions to be asked. How can the real become embryonated in the form of signs? Under what conditions do some of them, becoming mathematical, attain the real better and even show it? What status should we give to those that only attain it in profile?

On God

Leibniz called this stock divine understanding. At the limit of all our expertise, we can understand this understanding, even hope to make its parts and lodging. Does God, whose acts were described by Leibniz's seventeenth-century philosophy, draw from this table of notes, growing like the combinatorial explosion, in order to bring this world into existence?

The one Pascal called the God of philosophers and scholars therefore personalized a global sum of information of this type, or its equivalent as dreamed of in the past, from which we can now draw, at least virtually, spaces, times, the Universe and existences, proofs and narratives. The inventor of the calculation of the division of the stakes nevertheless repudiated this God, for excessive abstraction and no doubt deductive preformation, in favour of the God of Abraham, Isaac, Jacob and Jesus Christ, a God that would be contingent and historical and better attached to the probable and improbable events of human existence. He was right all the more at the time because the dominant mechanical model was causing the radical contingency implied by creationism to be forgotten, and it is so often forgotten.

I am humbly tempted to reconsider Pascal's decision and call God, once again, the prosopopoeia of this final place, broad and temporarily global, because this vast habitat includes both necessary deductiveness and contingent probabilities, better, the four modes: the necessary, the possible, the impossible and the contingent, but also the universal and the local, the heteronomous and the autonomous. We are partially opening the door of this Understanding.

4 THE GREATEST CONTEMPORARY DISCOVERY

Consequently, if I had to choose the greatest invention of the past century, I would decide in favour of the following one.

The Galilean model

At the end of the Renaissance and at the beginning of the seventeenth century, Galileo announced that nature was written in mathematical language. In opening modernity, this discovery transformed our view of the world and our way of dominating it (through the new technologies based at first on mechanical models), changed our society (become productive in a new way) and lastly our style of thinking.

Stemming from science and decisive for philosophy, this remark confirmed the separation, opened up by Plato upon the discovery of Greek mathematics, between a heaven of theoretical ideas and the apparent world of experience. In achieving a new union between these two domains, Galilean science no doubt reduces their distance but doesn't cancel it out. Mathematics describes or explains phenomena, but from the outside; *episteme* ensures theoretical knowledge, and *techne* does nothing but apply it. In addition, this rupture and this union assume three traditions that are at the root of the West: the Greek rupture, as I just mentioned; the Jewish union, insofar as Yahweh, transcendent, incessantly realizes himself in the immanent history of the chosen people; more still, the Christian incarnation since the Son of God unites in his flesh the two persons, divine and human. This is a prosopopoeia, again, of the modern tangency of the heaven of geometry and the earth of appearances. By incarnating mathematics, physics was just born from a culture, even against it, that conditioned its emergence.

The Galilean exception

But the Galilean discovery implied an exception. Only the inert, mechanical and physical world – the motion of the planets, the fall of bodies or light rays – was written in mathematical language. Nothing else, neither the bowels of matter, already long explored by a timid beginning of chemistry, nor the life of planets, animals and humans, a life having always been observed with a delightful precision, entered into this result. This vast remainder let us think that the sciences said to be hard didn't exercise their effectiveness on gigantic sections of existence, consequently approached in a thousand other ways.

One can conceive the frustration felt for centuries by inventors of elementary symbols, such as Lavoisier or Mendeleev, by naturalists, such as Buffon or Claude Bernard, nevertheless close enough to algebra to discover genuine theorems there and yet excluded from its exactitude. One can imagine the bitterness of those who were expelled from the paradise of the sciences, thenceforth abandoned to a knowledge said to be less 'hard'. One can try to count the theories and behaviours stemming from this resentment. But why this exception? Nothing forced us into it. Hence the scandal.

The end of the separation

We have just filled this ditch. For, ever since we learned that DNA winds into two helices and that it constructs proteins through the intermediary of an exquisite wire drawing machine programmed by genes coded by means of a simple alphabet, and since we have been deciphering this sequence for every living thing, we have finally been including these abandoned sections.

Yes, life, having a chemical nature, is written in mathematical language as well. But the Galilean model is no longer applied in the same way, and, on the occasion of this change, we have to transform our ancient formats of thinking. For life, 'written', 'writes itself', in addition. Formed from letters, from numbers or from notes in long sequences, the multiply folded strips of acids and proteins are transcribed and translated of themselves, without our intervention. The mathematical language doesn't have to be discovered; it lies in the very bowels of the thing. It neither describes nor explains these new objects from the outside, rather, present in their inner reaches, it constitutes them. From transcription to translation, from coding to coding, an initial sequence ends up making proteins and new cells. Theory constructs the thing itself.

On the other hand, the language written here cannot be taken to be mathematical in the Galilean sense, which used geometry or algebra in the

Greek style. This language unfolds algorithms, another side of mathematics, a linguistic one. 'Life writes itself in algorithmic language' – this is what a new Galileo would say first about contemporary invention. Not only 'writes itself', but 'constructs itself by writing itself'. 'Life writes itself in algorithms and constructs its elements by varying on this language.' Of course, life conforms to the line of physical nature but by slightly shifting its formal model, from geometry to combinatorial analysis, and by carrying out of itself what we did with this nature, through technological applications, after Galileo's discovery. We can directly read theory or language on DNA, the first element of reproduction, which moreover realizes their practice, as well as on proteins, endowed with their topological folds. We are beginning to bridge the separated domains.

The sudden identity of the algorithmic formal and the real formed by it and constructed by its variations therefore changes our modalities of thinking. There no longer exists an ideal heaven being reflected, from a distance, in images or shadows; there no longer exist visible appearances and intelligible realities, phenomenology, on the one side, and ontology, on the other ... In short, the dualism that had sustained our thought since the Greek dawn has dissolved. What passed for a miracle in the eyes of Kant and Einstein, this mystery of the successful harmony, since Galileo, between mathematics and experience, is something we don't even have to think about; life continuously achieves it, for life reproduces and develops from this union without any distance. This singular algorithmic language of itself constructs living beings; naturing, it causes them to be born and, in unfolding itself, launches them into time.

Can we, additionally, dream? Superstring theory will no doubt soon explain to us how the inert world exists from its links between quantum mechanics (whose laws regulate its smaller theatre) and general relativity (which legislates over the big one). Can we imagine, as though in parallel, a mathematics whose theorems would explain how, in the smaller theatre, the living thing exists from its links between the algorithms hidden in small cell compartments, whose combinatorial laws regulate their reproduction programmes; but also the elements of tensor calculus legislating, at large scale, over the behaviour of muscles and joints, the geometry and mechanics of the solid skeleton, the various topologies of the tissues of the embryo and of the formed organism, the hydrodynamics of its circulation ...?

Et verbum caro factum est

But before the dream and returning to effective discoveries, it is indeed a matter of the greatest contemporary discovery since, by giving mathematics the totality of the existent and of the knowable and no longer only one region, by noting, moreover, its effectively operational and constructive functioning, this discovery brings the era of separation between abstraction and the concrete to a close and therefore compels us to rethink philosophy, of language as well as of existence. It no longer fabricates the same machines, as we have seen; it doesn't point towards the same metaphysics.

In a celebrated article, Kojève once wrote that Galileo's discovery took place in southern Europe owing to Christianity, the central dogma of which expresses precisely the union of God and man through Christ incarnated in the double body, divine and human, a theological synthesis that to his eyes enveloped the Galilean synthesis of ideal formality and real phenomena. Today we are moving from this incarnation to what could be called carnation: for what would Alexandre Kojève think of the discovery that the Word, in the most exactly elementary sense of a sequence of letters, becomes, in the most concretely constructive sense, flesh for every living thing universally? That life is made of language?

Old technologies and new information technologies[1]

So let's now go into exemplary detail for machines and continue the demonstration from just now. How are we to describe them? A lever or a winch is used to lift weights, a car or a plane to move bodies, a cannon and a bomb to kill people. This is the pragmatism of *Homo faber*, who constructs tools for their utility, who perfects machines with a view to a finality. Modern science, since Galileo, has made these fabrications all the more effective because their geometric model grew in precision and exactness. So we now believe that every machine comes from an 'application' based on a physico-mathematical 'theory' and that they similarly have a use. False.

False, for the cell's machinery or wire drawing machine can no longer be called machines in the sense of the tradition, but rather, like Turing's machine, they are universal, theoretical and practical. They mix symbol and technology. False, for, in the sense of this tradition, a computer, a silicon chip, the internet …, the new technologies, every machine that's the child of Turning's, have no use. A new objection: housewives use the first in the list, for example, to draw up and manage the household budget; astronomers use it for their theory of black holes, voyeurs to drool in front of their pornography, mathematicians for their equations or curves, businessmen to play the stock market, artists to compose music and children for games. Response: these technologies are used, of course, for all these things, and even for a thousand more, precisely because they have no use. Dedifferentiated, universal, they transfer the builder's utility project to the utilizer, who employs them at leisure and as he or she sees fit. The one who builds and produces them cannot predict for what or to whom they might become useful or not. They have no direct finality. Without any initial intentionality, their function is discovered *a parte post*. Its builders don't yet know, for example, by whom or for what the 3G mobile phone will be used.

So contemplate now the vast difference between simple machines in the Greek and Cartesian sense or even those of the industrial revolutions, every machine in the sense of statics, of kinematics or thermodynamics, in sum, of force and energy, and the 'machines' (but why, good god, call them by the same name?) of the new technologies, equipped with their consoles, screens and keyboards. The first machines can descend from the theoretical heaven of mathematics to be 'used' by materializing. Alphabetical, algorithmic, the second machines form such a perfect synthesis between theory and application that no one can find the old separation again. We even have a little trouble distinguishing hardware and software here. Children of Pascal, Leibniz, Babbage or Turing, wouldn't it be better to name these machines with a word analogous to the words used for musical instruments: tablatures, for example?

Likewise, read some sentence written in your language and endowed with a meaning. Some sequence or other of letters, notes or keys, nonetheless used in a hundred languages analogous to your own or to some other, finds itself, most often and barring the exception of chance, devoid of meaning. Yet, starting from their alphabet, you can form, in one of these languages, as many sequences or sentences, lastly as many meanings as you like. Thus you fabricate signification starting from elements that don't have any. Likewise, with your computer, you obtain utility from a tool that doesn't have any. Hence the interest in putting elements coded in numbers or in languages in a table, better, in constructing this tablature by means of such elements. From this keyboard shoot up infinite varieties of utility, performance, meaning ..., music, as well.

Pragmatism and finality: Life and machines

This is why, in particular, the new technologies, also based on algorithms, so strongly resemble the things of life and allow us to understand life's development and evolution: for both of them use alphabets, coding, compressions, translation or programs, etc. Life knows as little finality as computers do because, far from resembling a machine made up of classical machines, which are 'used' in fact for single and precise functions, as Aristotle, Descartes and so many others have said, life is built like the new technologies, themselves algorithmic and without finality.

What is a hand used for? For nothing and everything: for breaking bread, for playing the piano, for hammering copper, for cutting the bad chasselas grapes from the bunch, for splicing rope, for throwing a ball, for hitting, for caressing, for taking, for giving ...; it ends up giving out signs and counting on its fingers. Continue on in this way from the most apparent organs – foot, elbow, brain or nerves – to the slenderest molecules. Yes, life has machines as its models, but this sentence has nothing to do with what Descartes or La Mettrie understood by that. Linguistic and algorithmic, our machines are undecidably theoretical and practical; they render the latter distinction archaic. We shall have to talk about totipotency.

For a living and algorithmic music

Regarding the genome, biochemists therefore talk about a text and judiciously compare its sequence of amino acids to a series of numbers or letters. Even better than a metaphor, this representation functions excellently. Has a comparison to music ever been tried? It nevertheless has an alphabet, made up of notes fewer in number than the alphabet of an ordinary language and even than decimal numeration; it strings them together into melodic sequences whose length – algorithmic complexity – depends on the techniques employed by the composer, fugues, ritornellos, counterpoint, da capo ..., which allow its 'compression'. It codes them according to various keys, F, G, C, etc., according to registers and ranges; it uses transcriptions of every kind. Better yet, it wouldn't exist without an objective and concrete medium: scores on paper, voices or instruments interpreting it; and, in this passage from the symbolic to the real, through this 'expansion' into existence, it becomes individualized since none of these interpretations resembles any other; therefore its coding cannot be separated from the singularity of its material and sonorous realization.

In their construction, these media appear like and resemble the machines of the new technologies and even certain apparatuses in living organisms: keyboards, tables, pipe organs, vibrating strings ... On each medium, any music can be endlessly played or composed; the comparison here seems more instructive than the comparison to bricolage. *Le Système de Leibniz* had already noted that musical instruments didn't evolve much because they enveloped an almost infinite combinatorial time that made all sorts of compositions possible on them. They launch other evolutions than their own. Had the vast difference between a machine in the traditional sense and a musical instrument already been noted? It instructs us about the difference we observe between the old technologies and the new ones.

Just as universal as life and the human phenomenon of talking or the living phenomenon of transmitting and receiving signals, music has less meaning than language: either it doesn't have any, or it runs in an infinitely wide range of indetermination or possibilities; consequently, this signification only occurs in the presence of a subject or a consciousness. Lastly, the relation of music to time compels us to observe that we don't have any experience of duration except through music and life: these latter two are linked by so many relations that their proximity never ceases to amaze. Is it a matter of such a similarity that it hides secrets? Can I, again, dream that DNA vibrates like a string and that its very form demonstrates this vibration?

Unpredictable, constructive, singular, inspired and fragile, ready to collapse but persisting in iterating their finitude, necessary sometimes and miraculously, possible often, contingent always, life and music unfold in the space they occupy and move forward through the time they give rise to, symbols taking control of the surrounding universe.

5 *EGO*: WHO SIGNS THESE PAGES?

This great contemporary discovery allows me, for the first time, to sign the preceding pages. How? On my identity card appears my name, whose archangelic first name denotes belonging to the subset of males who carry it and whose proper name denotes a second belonging to the homonymous tribes, scattered across the mountains, from Mongolia to Basque Country; this not being the mark of any great singularity, the alleged identity vanishes here, replaced by a sequence of belongingnesses: name, first name, sex, date and place of birth.

Since I belong, as well, to the class of people who use a Macintosh, my name could be written APPLE 6798453 or even more simply by the barcode 67503167562099, which would be enough for the police to locate me. This is a convention and a code-thing link in which meaning does not occur. To assess for what and to whom postmodernism and deconstruction can be of use, notice that this arbitrariness, close in fact to nonsense and play, suffices for the power that surveils or the finance that knows how to register a trademark to catch me in both cases. But this sequence of numbers or of notes doesn't constitute my true name.

Which is: write, if you can, the very number of my DNA, shorter than was believed not long ago, plus the set of combinations among the genes that contributed to constructing the hundreds of thousands of molecules each cell of mine consists of, to differentiating these same cells by families, and lastly to specifying the colour of my eyes, the height of my stature, a certain programme of health, an original relation to the environment, etc. This is my name. My signature would exceed in length the volume of this book, for the combinatorial explosion launched by these numbers gives it an extension that would fill several works or, in a line, a vast distance. It would surely take months, and maybe years, to count, to decipher my signature, to read my name. But, for the first time, it concerns my true name, the code that corresponds so well to my body that it constituted it and that no other code corresponds to it. Therefore blessed are all the living things that escape, by means of the naked and interminable truth, the police and high finance, like needles in a haystack.

Who am I? A singularity that signs with this astronomical figure, a dizzying well of almost inaccessible numbers. There is no arbitrary convention of coding between this number and my organism: the one says the other exactly and vice versa. So there is an exact relationship of the figure and the person or in general between words and things. The fact that it proves to be almost inaccessible doesn't prevent the fact that it exists and can be shown; better, this inaccessibility forever forbids being able to deconstruct or appropriate it.

Real objects, the rocks and water from wells, straws and haystacks, are sets of numerical wells or of numerical stacks sown with needles of gold. Yes, all things are numbers. The real comes from this exact coding. Therefore it is discovered and not invented. Therefore it has meaning. Therefore it constitutes the inalienable patrimony of the totality of humanity. Since it existed, exactly, even before it was discovered, no one could invent or appropriate it.

Therefore I am signing true, although virtually.

PART TWO

THE WORLD

The Argument
The Greatest Contemporary Event
Old and New Common Houses
The Evolutionary House
The Second Loop of Hominescence
Who, Ego?

6 THE GREATEST CONTEMPORARY EVENT

During the 1900s, the percentage of families devoted to agriculture and livestock farming exceeded half the population of the Western countries. It has lowered, in France, from 8.3 per cent in 1980 to 3.3 per cent in the year 2000; agriculture's share in the gross domestic product shifted, during the same period, from 4 per cent to 2.3 per cent. In the majority since the Neolithic, the date of its invention thousands of years ago, agrarian activity lost almost its entire population in one hundred and fifty years, with a notable fall from the 1950s onwards. The First World War had already put to death so many peasant-infantrymen, whose names crowd the stelae of Europe's villages, that one might even think that, implacable, it was pursuing this goal. One now has to go to China, the Philippines or the Andes to see groups as dense as during my youth working in the fields.

Up until the aforementioned dates, practices and sciences, arts and religions, languages and cultures flourished with and due to agriculture; a blacksmith, a shopkeeper, a military man, a priest, a judge, a lawyer, a politician, even an inventor or an economist … in the city had a direct experience of what, more than a profession, provided the very model of human activity and of its relationship to the world, and which still remains so for a billion and a half people. Our decision-makers today and those they administer have lost every link to the earth, to the living things of flora or fauna; their world view has lost the world. They are unaware of how the *paysage* [landscape] is sculpted, that this word, like the word *paix* [peace], comes from *paysan* [peasant]; they have never seen wheat or vines ripen, calves, pigs or clutches be born; therefore they can no longer read *Mr. Seguin's Goat*;[1] my students have never tethered one to a stake. Last year, the chief executive officer of a large European milk enterprise was shocked to learn from my mouth that cows were milked after they gave birth; he and his wife were unaware of this. More than a century ago, on the other hand, Pasteur was only occupied with agrarian questions: tartrates, beer, wine and milk ferments, animal rabies … The popularity of his work was, of course, due to the fight against microbes but particularly to the fact that it occupied a decisive place between biology with agricultural import and the biology we know, without earth or life. For, ever since Buffon, Pliny and Aristotle, natural history came from the peasantry and returned to it to improve its practices. Except when he might work in agribusiness, what biologist today is concerned with, if not wheat or pigs, but even life?

This is the greatest event of the twentieth century: the end of agriculture, insofar as it shaped behaviour and cultures, sciences, social life, bodies and religions. Of course, farmers always feed and no doubt will feed their contemporaries; but their acts and their existence no longer steer humanity, no longer result in a humanism, no longer give them frameworks of space and time. The West has just changed worlds. The Earth, in the sense of the planet photographed in its globality by cosmonauts, takes the place of the earth, in the sense of the plot of land worked every day. This crevasse separates the end of the past century from all the time that has passed since the Neolithic; it has already transformed our relations to fauna, to flora, to seasonal duration, to time, to the weather, to inclement weather, to space and to its places, to the habitat and to our travels. It has changed the social bond; we no longer live together in the same way as soon as our common nutritive cord to the field, to the soil, to the grass and the animals, to the occupation of places, to their defence and to war is erased; we don't die in the same way since, preferring to burn our dead for lack of space in the city, we no longer bury them under the ground irrigated with the sweat of our labours. Who today knows how to read 'The Farmer and His Sons'? So this is one of the broadest and deepest new things of the past century. Consequently, regarding the landscape, ecologists from the city, little familiar with the old rural world, say poorly with words and only recently what farmers have done there with their arms for millennia. Cut off from the earth or ignoring it to the point of contempt, philosophy, at least to my knowledge, has not yet taken this hominescent rupture into account.

From the body to agriculture and to biotechnologies

I described the change of bodies first because it precedes a second agricultural rupture, which is beginning now, and whose newness extends the one I just designated so as to plunge with it into an even more distant past. For the new body – victorious, at least for a time, over its latest predators or parasites, microbes – required, beyond health, food security and a certain control over reproduction. So in the recent biotechnologies, the two ruptures, the agricultural one and the corporal state one, flow into each other. The advent of this confluence refers to long eras.

The first regular crops appeared at the end of the last ice age and didn't result, as has long been believed, from a slow spread starting from the Middle East since corn appeared at about the same time in South America, a species selected starting from a mutant of a wild plant, teosinte; outside European influence, this invention dates back 5,000 to 7,000 years. A plant monster, wheat combines the chromosomes of three species of cereals: how our ancestors obtained this now-necessary and common GMO, we don't know. The first domesticated animal, the pig, saw the light of day in Turkey more than 10 million years ago; the goat and the sheep followed. A 5,400-year-old Sumerian text describes grain fermentation and beer brewing; 6,000 years ago, the Egyptians and Chinese made bread, cheeses and wine, which presuppose the knowledge of yeasts, at least in practice. These cross-breedings and technologies are already biotechnologies, which try to master the genome, invisible, and microbiology, unknown, starting from visible phenotypes. So in their struggle against bacteria, Jenner, Semmelweis and Pasteur repeated an ancient tradition that the Bible has begin with Noah, the first viticulturist, therefore the master of certain ferments. Better yet: examine our agrarian landscapes and go back up from our contemporary ones all the way back to the night of these times and confess that you only see around you vegetables, fruit trees, bushes, even forests … planted by humans and inhabited by animals hunted or domesticated by them, consequently those biotechnological varieties I have already called PMOs. You have to hang nets in the canopies of tropical forests to attain living beings free of all human intervention. But if we protect them in parks, they enter into technology and culture. The end of agriculture as a hominian model is accompanied by the general humanization of every species: the beginning of a 'cosmoculture'.

Therefore, in Ghent, Marc Van Montagu and Jozef Schell obtained the first transgenic plant in 1983. Their technology improved and developed everywhere without any media coverage or public anxiety for fifteen years. The increase in power of these new biotechnologies poses a series of questions, the most important of which today concerns less genetic engineering itself than the monopolization by multinational companies of the seeds so modified. Agrarian, the first rupture overturned the twentieth century. The same is true for the rupture of our bodies. Scientific and laboratory, the latest rupture affects the present and our future. Coherent, these ruptures together concern all the species, our own included, and the world: the world of practice, life and knowledge. These ruptures follow one another and bring to a close an epoch of at least 10,000 years. If they transform the relation to food and to reproduction, to oneself and to others, to the landscape, to space and to time, don't they change humans? Produced in certain collectives, in the course of their history, by their sciences and their technologies, in their economy and their politics, these ruptures affect, beneath these cultural components, the 'nature' of humans and of the world. That is why I call such ruptures hominescent.

The first scenario

The distant past therefore shows us the continuity beneath these ruptures; but this coherence only appears because we return backwards or through a kind of retrograde movement. Since time advances chaotically and contingently, this reason doesn't allow predicting or even sketching out the scenarios to come. Which ones, however?

First, are we heading towards the triumph of the biotechnologies? Tomorrow, will there be two kinds of living beings, those genetic engineering will have recombined, whose patents have already made, and will continue to make, the fortunes of the proprietors of vast industrial livestock farms, and those that will subsist in parks and museums? Connected to the surrounding countryside and the tropical forests, will zoos and botanical gardens join with fossil rooms in order to group together, in the same wide space, the farming species of the ancient domesticity, plus those species whose archaic wildness will be reorganized by unforeseen laws, lastly the species that vanished during the antediluvian eras, joined all together on display for tourists and school children to visit? Will we then see behind the same wire fencing, footlights or glass pipes in the sea, fossil flora and fauna, wild and domesticated, mixed guests stemming from a vanished time dating at least from the Cambrian explosion and ending this morning? A more than million-year-old rupture. Will we therefore distinguish the living things that are planned, fabricated by human hands, transgenic, patented, consumable, available on the market, adapted to our needs, harmless and sometimes cloned … from those of yesterday, evolving in a nature finally totally subjected to culture? Will we therefore divide the natural into two artificials: the one, scientific and technological, intended for industries, whether food industries or other ones, ensuring our healthy survival, and the other, juridical, applied to what is left, for the pleasure of our vacations: lab-city mice and zoo-country mice?

This science fiction vision is appearing on our horizon; every day we see several steps being taken towards it, including under the influence those who reject it and seek to protect the species, rendering them cultural by this very fact. Why wouldn't the biotechnologies invent new varieties with this or that characteristic, for survival and ripening, transportation, sale, how long they keep, and taste since, as early as the seventeenth century, for example, we invented the nectarine, a cross-breeding of the plum and the peach, resulting in a hardier skin than this latter, therefore easier to market?[2] Newness encounters the tradition more often than we think. Since the first cross-breedings date back to the Neolithic, GMOs to 1983, why wouldn't the archaic trend, strengthened by the more recent one, continue on? It follows, but for a bifurcation, the straight current of the retrograde movement I spoke about at the start and even the straight current of evolution, if not in Darwin's sense, at least in the sense we give it today. The chairperson, from 1988 to 1997, of the *Commission française de génie biomoléculaire* [French biomolecular engineering committee], Axel Kahn observes that no article in any newspaper nor any protests took place before 1996, a date that followed the epizootic in Great Britain called mad cow disease. According to him, if a battle takes place it will start too late, for the change of direction has already had multiple effects. We cultivate corn that's resistant to the European corn borer, soybeans that no longer need herbicide, tomatoes and melons that are more solid, lastly fruits containing more terpenoids, 'natural' aromas. And how are we to control from now on the mixing of seeds flying in the wind?

Several variables play a part here: resistance to disease, productivity, control of maturation, the environment and safety. Regarding the latter two points, who isn't aware that a wild and old living thing will easily prevail over another one, recent and 'artificial', therefore infinitely more fragile? Therefore the dangers lower with science, not the reverse; the news items announced in the media do not alter the stubbornness of the facts: food poisonings are continually being reduced; many medicines today derive from GMOs. And we continually learn at our own expense what we risk by trying ignorance. What remains to be learned is if and how research is hand in glove with the financial powers that are taking the place of the scientists in the laboratories as they have for the peasants in the occupation of the land. The question is asked with even more acuteness for the third world than for the West.

The second and third routes

The second scenario: Can one imagine a trend reversal owing to several ecologically inspired reactions opposed to this gigantic bifurcation and its consequences? Perhaps desirable, this movement has several components: geopolitical, economic, cultural … Western, the battle sets Europe against the United States, whose sensibilities differ owing to the antiquity, in the one, of the agricultural tradition, and, in the other, its recent and quickly industrialized character. On the other hand, in France, as in Morocco or China, rare places, what matters to everyone is the taste of a dish; in the United States, a strange vocation for obesity doesn't distinguish the quality of a restaurant so much as its price and maintains adults in the state of infants afflicted with ageusia and incessantly sucking at the teat.

The last possibility: by an improbable miracle, the developed countries care about the third world and seek to restore the balance of humanity's food, in which one person, today, at this very moment, dies from hunger every three minutes and four out of five complain of not having enough to eat in countries exporting foodstuffs to chubby Westerners suffering from a lack of diets. Supposing that Africans, South Americans and Indians might someday drink and eat as generously as obese Westerners, agriculture and livestock farming would again become a major issue. Without our always realizing it, the first two scenarios adorn the utopias of the rich countries. What would happen if the 'democracies' agreed to abolish an appalling inequality and to no longer kill the children of their brothers by means of hunger? Would we return then, for an incalculable time, to the era of agriculture and domestication?

Here is where the question of biotechnologies resides: Do they develop varieties intended to remedy the problem of hunger? If yes, then who would reject them? Here is where everything is decided. And three examples won me over to their side: the invention of a modified rice, called golden rice, supplemented with vitamin A and iron; 400 million people in the world are gradually becoming blind through vitamin A deficiency, and a billion are suffering from undernourishment through lack of iron; 300 grams of golden rice per day would make up for these deficiencies. Who knows this piece of good news? Except for various defamations, it has had little mention in the media, nor has the news of the discovery of two other GMO plants, one of which is adapted to alkaline soils, which make up 30 per cent of the world's fallow land, and the other one of which is adapted to saline soils, which are thus given over to cultivation. Supposing that these varieties might spread, while observing basic precautions, who then would forbid the extension of agriculture on to soils that were unsuitable up until then; who would oppose the blind seeing and the hungry eating?

The oldest citizens of the world: The peasant, the scientist

Here as is often the case, the benefits counterbalance the dangers. Good and ill are rarely distinguished the way the devil and the good Lord are. Agriculture has been confronted with the global already for millennia. Departing from two rare and very remote from each other sources, the Fertile Crescent and South America, agriculture invaded, by fragments, the planet and has therefore been globalized since the Neolithic. The first big planetary tours mixed food staples, then spices; I'm not counting microbes. Compare, for example, Petronius's mixed table with the frugal table from Plato's *Banquet*, olives, almonds, wine and honey. Europe in large part resolved its endemic famines by exchanging fruits and vegetables with the entire universe: it cultivates them, but neither the tomato nor the peach nor the potato comes from it, nor the turkey, stemming from America, nor the guinea fowl, from Africa. Rarely scraping the earth, the Australian Aborigines surprised the English and Quebec convicts, who believed in the universality of *Homo rusticus*. As a result, the livestock breeder, the farmer, the forester find themselves to be much less déraciné when they travel than anyone else; cultural manners distinguish us, cultivating ways bring us closer. Yes, the Aborigines made me feel more déraciné, because they themselves were 'déraciné' in the literal sense,[3] than the Indonesians or the Chinese, despite the ravishing difference separating horizontal terraces of rice and waving plains of wheat. Today still, the trades of the land have achieved the greatest progress in productivity, therefore first from the origin and still first in the contemporary era. Globalized thus from the dawn, can they today *not* take the third world into account? Is it dying of hunger because of our actions? Is it suffering this punishment through some deliberate will on the part of the invisible hand? How is the death penalty this iron grip has weighing on them to be abolished?

As for science, it likewise became globalized, later of course. Food and knowledge achieved the status of general equivalent before money, which appeared later: bartering cattle preceded currency, as witness of this, the word 'pecuniary', which we use for it, even though the word signifies the herd pushed forward by the herdsman. Even before Pytheas the Massaliot, the expert explorer of the fifth century BC, the first great travellers departed to collect plants, to find other species of flora and fauna, to know them in order to raise or sow them, before wanting to observe the stars from another angle. Alexander's generals sent plant and animal specimens from Asia to Theophrastus and Aristotle as though expeditions of war took not only booty of jewels, but also and above all booty of plants and animals. And how are we to describe botanical gardens if not by means of their globalization? At least since the seventeenth century, walks there traversed a global herbarium, named in that universal language invented by Linnaeus. Neither Pliny nor Buffon contented himself with describing the bestiary around Lake Como or the hills of Burgundy. Thus the peasant and the scientist, old citizens of the world, anciently associated, can and must understand each other to free themselves from the yoke the recent masters of money have weighed them down with.

The new citizens of the world: The peasant and the scientist

The contemporary state breaks, by exacerbating them, trends that have always been present in the global history of agriculture. Every advance in the productivity, discovery and proliferation of cultivable and domesticated species resulted in a corresponding increase in demography that plunged the peoples delivered from hunger by the first steps into a second famine, one due to the number of consumers outstripping the volume of production, which therefore demanded a second effort. Over this helix that's always being repeated, by means of deviations from equilibrium, soared ferocious dominant beasts, the transhistorical parasites of the peasantry, profiting from these gaps and from those induced by variations of the climate – thin and fat cows. Hence the dash forward into the world's space in search of cultivable lands, where the harvests, always insufficient, would pose anew a similar problem. Conquests always proceed from the poor and from hunger, from weakness more than from strength.

Did agriculture become generalized in space and time by this race itself, by this ever-recommencing gap between *exploitation*,[4] taken in both senses, farming and human, and demography? Invariant across its variations, this impetus, inevitable and often pathetic, is today reaching a global apex, even though its conditions, now global, no longer present many margins for other headlong rushes into either the expanse or time except by demolishing the walls of several last great domains, by winning back infertile soils, by finishing cutting down the forests, a dangerous enterprise. At these relatively precise limits of spatial density and ecological risks in time, how can we not move to the scientific, to the virtual, to the numerical, I mean to expansions that no longer happen in the soil or in the landscape but in laboratories and at another scale? Can we really avoid this leap, truly hominescent, into knowledge? Elsewhere and differently, we have already accepted it ten times.

The fights will take place here: Should we run quickly and forwards towards large deviations from life such as we knew it, towards biotechnological engineering, without giving in to panic before the scientific innovations, or, more slowly, towards practices said to be gentle, 'organic' in the now popular sense? Which should we choose? Biochemistry in the hard sense or ecology in the soft sense, I mean in the sense given to it by politics? How can we escape from the parasites? Certainly not by fighting against science and returning to the oldest traditions since at the beginning of the last century, France, for example, one of the most fertile agrarian countries in the world, had more farmers without however feeding itself, while it does so and exports today. In addition, where will the fear unleashed in the public lead us? I believe fear to be a less good councillor than discovery. So this is my utopia: that the scientists of the world share their expertise with the peasants of the world; that these mixed groups protect themselves from their parasites; the sciences of the earth to the men of the earth. Otherwise the planet will have to renounce its common name of Earth and adorn itself with the proper name of a seed company for advertising purposes.

Will the three scenarios combine in the end for an agriculture that would in sum be well reasoned? Battles always set pugnacious radicals in opposition, but time and reality end up uniting them. More and more science is needed to feed the people condemned to death in the third world, a result that adds the first scenario to the third one, but the evolution of customs is instead going in the direction of taste, which favours the second one. We are already living today in a mixture of the three.

*

What can be said to be contemporary?

Such ill-advised scenarios, as I said at the beginning, are quickly refuted by this or that unpredictable event suddenly imposing a contingent solution: therefore it would be better to examine our daily practices: I find them, likewise, to be archaic and new, rational and mythical.

For during these days marking the beginning of the third millennium of the Christian era, Great Britain is lighting pyres on which 2 million animals are being burned: cows, pigs and sheep.[5] What human ever saw such hecatombs without nutritional purpose ever since humanity became humanity? After Abraham became the first to abolish the death penalty for the human species by saving his son Isaac, he made the knife and fire of our violence fall upon an innocent ram whose horns were mingled in the branches of a neighbouring bush. Did we, on that unforgettable date, start to eat meat? My language derives the word *boucher* [butcher] from this *bouc émissaire* [scapegoat] or from replacement. At least, human sacrifice gave way to the less savage rite in which bulls, calves, pigs, sheep or pigeons died. From that day on, individuals, families, tribes, small villages, even capitals sacrificed animals on solemn festivals and often to protect themselves from or treat themselves for the plague. During purification ceremonies, the Romans practised the *suovetaurilia*, in which they cut the throats of a sow, a ewe and a bull at the same time.

Holocaust during times of plague

We no longer content ourselves with a single lamb nor with immolating pigs, rams or white heifers on the occasion of crises or similar epidemics. Lighting up the horizon and filling space with their greasy and acrid smoke, the pyres flaming on the British Isles generalize to millions of animals the ancient hecatombs and holocausts, so named, in Greek, because either a hundred victims were killed, or they were burned to total destruction during the sacrifices offered to the gods in times of adversity. To whom are Europe and England sacrificing this immense mass never before sacrificed by humans? To the appetites of those who eat their fill? No, since these destructions don't change anything in their diet. To whom, then? To the gods being reborn from the present plague? To Dolly, goddess of the future, recently cloned? To the health of the ever so small number of extremely wealthy humanity? In reality, to the end of livestock farming, an end accompanied with peasant suicides.

The abandonment of monotheism is everywhere reviving the ancient machine for fabricating gods in large numbers. I have long known that we have returned to polytheism and even to human sacrifice since we tolerate the deaths of thousands of young people per year on the roads, sacrificed to the automobile machine-god, and since the bloody gods of television have already forced twenty thousand murders on a child of fifteen in order to inculcate killing into him. So from distant eras up to contemporary times, this type of sacrifice always took place in order to make these gods be born and proliferate, especially when some contagion was raging, except during the interlude of the era that forbade all sacrifice and is at risk of ending these days. But this has never happened in such a large amount, nor with this intensity. Two million bovines and ovines! Do you smell those charnel houses with their charred carcasses? What an aroma of antiquity! At the very moment when postmodern sensibility is tolerating animal pain less than human death, public opinion, administration, politicians, the police, the army, the media … (but who then is lacking from this lynching circle?), all of them assembled, all of them frightened, all of them counselled by ethical prudence and the exact sciences, all of them in complete agreement, all of them impelled by reason, are lighting these gigantic blazes. From precaution to holocaust: What strange Möbius strip connects, in this exemplary juncture, the most rigorous arguments and our most refined technologies with the most savage sacrificial rites, and the present times with archaic ages?

The golden calf in the temple

In one of these infernos, a few sensitive souls found, among the thousands of blackened carcasses, a young calf that had survived, from then on named Phoenix, reborn from these ashes; despite the microbes it was known to carry, the prime minister decided that the country would save it. Quickly becoming a media star, it was paid thousands of pounds to appear in public. Delivered, deified in the midst of sacrifices, crowned, made of money, you have recognized the golden calf. A clear-headed Moses would draw a parallel between this ghost of the apotheosis and the Roman woman who recently made a dead princess into a goddess.[6]

Who can doubt that our governments have turned against teaching the languages said to be dead, which would have preserved the memory of antiquity, so that we won't understand that this antiquity is returning? Our actions closest to rational decision produce ceremonies of such a gigantic archaism that we don't see it. And yet, this evident fact imposes itself all the more since, at the time of the shuttle *Challenger* explosion, the same sacrificial ambiance reigned not long ago at Cape Canaveral. What does this oblique weld between the ancient and the modern signify when it seals the encounter of the sciences, technologies and society? Overflown by orangish glimmers and pestilential clouds, Great Britain, the first formerly in industrial livestock farming, the first today in its ending, is becoming the advanced temple of the modern–archaic polytheism with a thousand blazing altars. What hellfires will we release when, after the epizootics, epidemics of antibiotic-resistant bacteria decimate humanity? Oh, new gods of our health, we pray to you!

Let myths reappear and here is an origin.

7 OLD AND NEW COMMON HOUSES

Before it died like this, and returning to primordial times, how did livestock farming proceed? What habitat does the word 'domestication' refer to? What relations have we maintained with animals ever since certain species entered into our house?

We imagine today that technologies only concern inert objects, tools or machines. But hunting, gathering, livestock farming and ploughing require expertise directed towards living things. Past existences, from generation to generation, in the company of domesticated animals, presuppose that humans constructed a common house with them. So this habitat in which days and works mix produces varied cultures in which life yields to our expertises and in which, conversely, our technical skills vary with life. As I have lived within one of these cultures before it died out, I can testify that it included an authentic teaching, through reciprocal breeding of humans and animals, and, better yet, constructed a genesis of knowledge in the sense philosophers give to these words. I am going to describe this age-old process, still inscribed in the memory of my body: this collective habitat, this culture and this cognitive.

The event whose importance the preceding pages have attempted to assess erased this memory and stopped this process by separating those who used to live together inside the ancient common house of domestication: humans in cities and animals distributed in distinct boarding schools according to our needs. Continuous no doubt, the transition between the two states didn't erupt suddenly and was being prepared over a long time.

Our existence now occupies another house, one our knowledge is slowly widening from the barnyard to the planet, and in which, by a new twist, we are working out different knowledge. We think we have thus lost the living thing because refined technologies have led us to this second state by constructing this second habitat. I'm going to show next, starting again from our body, that we have, on the contrary, completely recovered it. Our body sees and hears now all manner of waves, formerly invisible and inaudible; it flies, it can procreate at a distance. Whether it's a question of space or time, of perception, of motricity, of food or reproduction, our various functions gradually cover the entirety of the slots the other living things cut out and occupy in their environment. I call this wholly new body the Biosom; this other neologism designates a body – *soma* – in the process of encompassing the totality of life. By thus leaving the ancient farm, common to few living things, and by visiting the entirety of the places cut out by all of them, it gradually constructs a global house. Our world, in total, forms the sum of niches, at least virtually. So we attain a totipotency I will lastly set myself to define.

So here are two cultures needing to be defined: the ancient house in which the living thing was more technological than it was thought to be; the new one in which technology is more living than it is said to be.

I. The ancient domestication cultures

Henceforth without horses or scavenger birds, the cities and houses of the West drive away all animal life from their walls and avenues even though thousands of rats and billions of roaches invade them against their will; quite the contrary, India grazes its cows in its streets. In these quasi-laboratories, aseptic in principle, was born the idea that animals, mechanical, reduce to the functioning of their physiology and to their evolutionary reproduction. So how are we to explain the colours and cries, the postures, figures and movements of their expressive bodies, which their fellows and we formerly took cognizance of in order to construct a symbiosis in a common house?

Roosters and oxen

My peasant memories recall this vanished culture. We lived together, animals and humans, in a convivial site, the farmyard, night and day for four seasons. We knew one by one the oxen, chickens, turkeys and pigs, to which we gave a nickname suited to their particular character, proud or indifferent, lazy or quick-tempered, each male and female of them thus demanding to be treated differently: we didn't milk Miss Temperamental, thin, traversed with quivers, lashing whoever approached with blows from her tail and battering them with kicks, the way we did Blonde Rosette, flabby, fat and placid. Each animal seemed to know the temperament of its neighbour and to adapt its behaviour to it. Thus, the herd formed a shifting society with the birth of a calf or when the oldest cow went into heat; the fowl of the henhouse, of course, lived among themselves, cruel ducks and yellow chicks with their innocent beaks, but also with the horses of fragile health, standing, and the pigs, sprawling; the farmer's kindness or firmness brought about different global collectives according to the use of her hours and the unfolding of her trousseau.

The small farm therefore fit together several fuzzy subsets: the animals lived among themselves but neighboured other species, under the authority of the man, the lady and the children. These piled up collectives didn't merely evolve according to the physical constraints of the climate and the seasons, or by means of the necessities of reproduction and feeding themselves, but also by means of signs and representation: the size of the bull and the sculpture of its muscles, the sow's pink porcelain vulva, the rooster's haughty bearing, the beret of its crest and the chromatic range of its tail, the seductive gentleness of the heifers, the colour, black or milky, of the mare's coat made an impression on each of us, father, mother, child or animal, according to the days of the month and the individuals, submissive or powerful.

Leave the causes and effects of physiology or the machine-body, Descartes and Claude Bernard, aside for a moment, to enter, without prejudice, into the pure emotive–cosmetic Darwin didn't think it beneath him to speak of, and you will truly find there the origin of knowledge. For this farmyard is populated with attitudes and gestures, lustres and bearings, arrogance and nobility, squabbles and pettinesses, plumages and songs, in sum, with a singular perceptual cloud belonging to it alone: red, cream or blue, high or mediocre, threatening or tranquil, competing and imitative. Each animal parades around in glory, hides or steps into the background, watches the attitudes of the others and evaluates them, sniffs their stench, listens to their squawking, coos, lows, clucks, grunts, hurls its cock-a-doodle-doo, whinnies and reacts to what it hears. Listen, in your turn, with attention: each farm has its smells and stinks, its coloration and its music, varied like the hues on a pigeon's throat, dominated by the voice of its master and the barking of his dog, lastly surrounded by its inner mustinesses: solid and liquid manure, cow dung mixed with straw and hay, pigeon droppings, crushed seeds, the slimy mud everyone trudges in from autumn onwards. So each animal's body dresses after its fashion, dances, sings, stinks, presents itself, sumptuously ostentatious, so as to get itself known and feared, mimics or camouflages itself so as to disappear better; leaving thereby mechanism and biochemistry, at least a little, it thus enters into relations and, by means of these signs, all of them build a community beyond cellular, organic and functional wholes.

Of course, for example, this pigeon wing is used by it to fly; who can deny this? Of course, during the early moments of its evolutionary formation, this new limb, which led from reptiles to birds, contributed to the regulation of temperature in the intermediary animals; but since at least these two advantages existed, why wouldn't we discover other advantages and, in particular, that of presenting oneself to others? These feathers, these colours, these chatterings, these mustinesses which show, hide or mimic, why wouldn't they be used in addition for communication and not solely sexual communication? Sonorous, visual and odoriferous, these appearances enter into a network of exchanges and sharings, already one of knowledge.

Does there exist a better case in which perceptual play precedes and prepares a collective culture? There isn't anything in the farmyard that hasn't first been in the network of these displays. As far back as the seventeenth century, La Fontaine didn't content himself with relating a few exploits by deer, partridges or beavers in order to criticize the theory of animal-machines; he also recounted more than a hundred *Fables* in which various animal societies, through perceptible links, differentiated themselves.

For there are assuredly cultures of the farm – as there are elsewhere cultures of the jungle, no doubt, but not knowing them, I cannot faithfully bear witness to them – whose diversity, in each site, depends on the fact that the male owner chose, at the market, this piebald mare or that white bull with black patches that bellows pitifully, on the fact that the female owner likes to wear green or grey skirts, that she bought, in season, some pair of guinea fowl whose nocturnal coughing disturbs those who are asleep, that the daughter of the farm washes her clothes white or likes herself soiled. Under the coarse tyranny of a brutal miser, I have known starved-looking farmyards possessed of no more hope than a concentration camp, but I have also known others that were happy to slumber, scented, by the side of a kindly tenant farmer in love with his farmer wife: *Poulère* didn't resemble *Ricoy*; one's neighbour didn't reign over the same town or the same yard. Just as each hamlet or chateau, as each hive or anthill, homogeneous with regard to the species of its inhabitants, nevertheless bears witness to a civilization belonging to it alone, so farms, mixing, in a more complex manner, different species with human persons who are just as variable, change endlessly depending on the style and the original character of these assemblages, which all present a landscape of iridescent mixtures.

Reciprocal domestication in the ancient common house

In the common house built in this way and lived in every day, a reciprocal domestication is in fact practised. Humanity doesn't impose its traits on animals without receiving traits from all of them: the female farmer-heifer, a goose and hen, waddles along with a basket on her hip, a turkey hen, and the male tenant farmer, a rooster and billy goat, licks his pink chops, a pig, the way a horseman, elsewhere, puffs steam from his nostrils in the cold dawn, the way a hunter bares his boar fangs when he aims; the Australian Aboriginal who transforms into a kangaroo doesn't lie so far from us. In order to know, you need to imitate; in order to train an animal, you need to know; in order to get an animal to enter into the farmyard, you must yourself enter into counterfeiting the animal. This reciprocity puts all this group at ease, which as a result feels at home. Starting from the body, this acculturation is done via a body-to-body in which everyone shows and receives. The rooster and the pig enter the human's home on the condition that the human also knows how to enter the pig's or rooster's home. So the domestication question is shifted: it's less a matter of understanding how we began to tame certain animals, therefore of giving in once again to anthropocentrism, than of seeing how the common hotel was constructed, the hotel in which the host-animals ended up living in symbiosis, at least an apparent one, since they were lodged, looked after and fed by the parasite-human; in order to prepare this common site, it is enough for the parasite to become host and the hosts parasites, reciprocal domestication becoming then another name for symbiosis and this latter continuing the cultural genesis undertaken, body-to-body, next to every living thing.

Levels of domestication

Every language specifies all the words of the relationship wonderfully. A wild animal – tiger or crocodile – is subdued but doesn't reproduce in a circus or a zoo; a single individual, its master, maintains a relation with it for a long time, while it remains wild for every other human, who didn't dominate it. When another animal is tamed, this signifies that it enters into relation not only with one individual but with those close to him: it comes into his family and his private life, as favourite, as his property.[1] Just as, regarding the origin of knowledge, philosophers remain silent about the culture common to humans and animals, so, when they meditated on the origin of property, they didn't speak about the taming of animals, one of the very first appropriations however. As Buffon said, panthers are subdued rather than tamed; but it must be added, symmetrically, that hamsters are tamed rather than subdued. Therefore, the subdued animal doesn't procreate at all; the tamed animal can, but its young aren't born all tamed already. Only Orpheus, that is to say, music, had, it is said, the power to tame tigers and cougars. Additionally, the relation, in both these cases, only has to do with the individual animal, whereas domestication concerns the species: the young of the ewe or the cow, on the contrary, being born tamed, don't know how to return to the wild state: they are domesticated. I would gladly wager that they only reproduce on the condition I am describing: the emergence of a culture, and even of a knowledge, in this common house.

We must also distinguish a domesticated animal from another animal that is tamed; the donkey, jealous of the little dog, in La Fontaine, bore the consequences of this: the little dog gives its paw, the other one gives its hoof; if the donkey enters the living room, Martin the stick will chase him out. Hence several boundaries or circles of property can be drawn: the first circle (the very word 'property' comes from this priority) resides in one's own body [*corps propre*] (so I needed to talk first about the new conversions done to it); the second circle resides in the house in which one receives one's intimates, family and friends, including favourite animals, one's cousin and her private dog, even her tamed hamster; the third resides in this farmyard in which the donkey kicks and brays amid the domestic animals; the fourth resides in the entire farm, including the woods where the hare, deer and boar are hunted, for the farmer will or won't give the right to hunt on his lands; but it will be said that the partridge here is less wild than the Bengal tiger, far away. The English language calls the rare animals that return to being wild 'rogue', and French calls them *marrons*: the mustangs of the American West, dingoes, the camels of the Australian bush. Supposing they revert to this savage state, then it is fitting to train them; the entire process begins again. These successive circles, for their part too, draw a genesis – spatial and progressive – of culture and knowledge.

All these relations presuppose that man and beast enter into the same human and animal site, wherever it may lie and whatever it may consist of, for all life, as I've shown elsewhere, presupposes a niche or a shelter, a delimitation to the ground, in brief, a border. In the variable nature of this enclosure, you can recognize the set of passages from being wild to being subdued, to being tamed, to being domesticated. Who would dwell in this private habitat, without recognizing themselves in it, of taste and smell, of calls and profiles, whose exchanges construct this site? If reciprocal domestication presupposes this common house, it is fitting to mix there fragrances and noises, heats, attitudes and displays. In this integrative mixture, you can recognize the emergence of culture, indeed, the origin of knowledge, which, of course, starts from the body but above all from the body-to-bodies between every living thing. Thus fables and myths, most often stemming from a fabulous past, recount, after their fashion and at least, this cultural dawn.

Subsequently, knowledge will forget it. First because city mouse is no longer acquainted with animals or the common adventure they have nevertheless been on for hundreds of thousands of years with humans; but also for another reason, similarly corporal. Just as our bodies, endowed for the aforementioned displays, to the point that Aristotle rightly called us the most mimetic of all living things, easily associate with animals in the common houses of domestication, so, also endowed to exteriorize their functions, they lose them by objectivizing them; they make them into manmade objects. As a result, surrounded more and more by technological objects, our bodies recognize in them functions that are or were their own and which, as a result, intensify their solitude, literally mental, for knowledge now comes less from the body-to-body taken on by the whole of living beings, an archaic and now-devalued knowledge, than from this function-to-function reflected by the new objects. The city resembles the shed where the tools are stored, a shed stripped of the smelly and colourful lives of the barnyard. The common house becomes the house of tools and politics. Life loses the game to the profit of mechanics and the mind. Better: there was, formerly, a mind-life, whose place this object-oriented mind is increasingly robbing. Life dies ... This sentence will have to be understood.

How nature enters into cultures

But, formerly or recently, before the Neolithic ended or before I advanced into age, between nature (which no one any longer truly knows where in the world to find) and cultures (prejudged to be merely human, mores, customs and fine arts), an in-between habitat existed (which the technological mind of the city has caused to be forgotten but from which we all came) which placed agriculture and livestock farming in common, of which farmyards offered so many varieties before they themselves enclosed battery-reared animals in separate stalls like human young in middle school, adolescents on campuses or the aged in old folk's homes. Clear and distinct, this mind detests mixture. Nature-culture, ploughing and pasturage, I don't know what to call the houses of yesteryear; choose at leisure, the title matters little, but I wouldn't like you to call what was being woven there subculture under the empty pretext that you didn't know about it.

Starting from this common habitat, here is its true name: domestication, which signifies above all that humans share the same house with animals. So don't seek first how or when humans enslaved certain animals, but what space which species haunt and share. Since certain insects rear other insects in order to suckle them, eat them, borrow their energy, even inhabit their warmth and since being a parasite goes with a large part of living actions, the crux of the problem of domestication indeed revolves around the house, not humanity or its actions. What goes on in this house of hosts? How, for example, in some farmyard, do homologous signs circulate? For, in this mixed society, animal and human, exchanges emerge that are already signalling and perceptual, in which a certain warm, common, reciprocal and, all in all, true knowledge begins, in which the strategies of hunting and camouflage, warnings intended to frighten predators, become diminished. Yes, animals recognize each other, individual by individual, according to the presentation of their bodies, size, colour, bearing and cries, scents, habits and movements, therefore already representation; the various species, among themselves, locate each other, revolt or tolerate each other, classify themselves by climbing the chicken coop, by taking refuge in the hutch, in the regime of the barn or the stable, their respective niches built by the farmer; all of them together, therefore, have a global relationship with the boss lady who feeds them, with the children who talk to them and with the dogs that protect them from the fox, whereas each animal maintains an elective relationship with each of the others. These shared perceptions form a beginning of knowledge. Certainly, the farm differing from the forest, humans compose, through hunting, agriculture or livestock farming, an environment that's ever new for animals, the way architects, scientists and artists construct another one for humans themselves, but each original environment above all depends on the appearances, forms and secondary qualities presented by the animals or humans so assembled.

These cultures fit into each other and are distributed across each other, pile one on top of the other or neighbour each other in successive circles. Taking refuge in its niche, each species lives apart but immediately swells up with an interspecific life in the barnyard in such a way that all of them together construct an original collective separated from life in the wild. Dominated by humans, whose faces and bearings aren't afraid to metamorphose sometimes into muzzles or snouts, into animal appearances and displays, these successive circles extend, little by little, to the steppe, the jungle, the mangroves and the forests through hunting first, then in the course of what could be called the second domestication, in which technologies and the universal expansion of the city affect every wild species, placed between extinction and protection, through the second wave of putting into cages when the Universe, recently, became a knowledge farm.

La Fontaine's source

Corporal and perceptual, the links animals maintain among themselves within their own species are found almost unchanged in human gatherings, in which everyone locates himself from the clothes of the others, rich or poor, cutting-edge or outmoded, or from the accents, Parisian, foreign or provincial, jointly erecting hierarchies following these simple signals; understand these collectives via reduction to barnyards. This or that feather mimics the rooster's tail, and this or that social precedence imitates the line of pigs in front of the trough. Conversely, the farm teaches us that animals establish communities together and with us by means of corporal signals and teaches us as well that culture and knowledge are born from the body, from theirs and from ours. The fact that we have forgotten these communities, that we only remember them, sometimes, across the convoluted language of some phenomenologists in no way conflicts with their rustic and quite immediate complexity.

So claiming that the Lion who takes the four shares of the Deer from the Heifer, the Goat and the Sheep copies the actions of some Sun-King, while the Wolf, the Fox, the Bear and the Tiger represent aristocrats, and that the Donkey and the Hare obey like serfs and lackeys makes the peasants we were and remain laugh open-jawed. First of all, Latin society would have to resemble the society of the French classical age like a twin, and this latter would have to resemble the collectives of earliest Greek or Hindu antiquity for the same schemas to return from Aesop to Phaedrus and from Pilpay to Benserade. How many times then did the Lion and the others change their names, images, models, masks? Let's pass over this social history, one flaccid enough to only fascinate French critics, intoxicated, as we know, with politics.

But above all, this is a misinterpretation: the *Fables* represent our human societies so little that, to the contrary, our societies repeat and continuously imitate animal groups. These little poems don't show the city or the Court but directly the barnyard or the clearing, whose customs and habits are reflected wonderfully in tribunals, public gatherings, plazas and fairs, salons and palaces. The application, reading and translation doesn't go from human collectives to animals but in the other direction, from the latter to the former. La Fontaine's source, but also Phaedrus's, Aesop's and every fabulist's source? Here it is, simply, flowing in the drinking trough common to horses and oxen, to ducks and dogs, placed right in the middle of the stable and the hay-barn, flowing in the forest watering hole, where the deer intimidated by the wolves and boars drink. So in the king or some little leader, recognize the rooster, its feathers and vanity, or the rogue tenant farmer; on the kowtowing spines of courtiers, read the obsequious backsides of hens; associate the bright tints of the skirts worn by the Marchioness, headily perfumed with signed chemical syntheses, with the female guinea fowl, the turkey hen or peahen, which are surrounded far afield with the powerful scents of substances that attract males; men and women give out signs by means of their bodies neither better nor worse than these birds; representation by means of gestures and attitudes, clothes, colours, accents and smells repeats, without any great change, the cries and chatterings, the feathers and bristles, the bearings of the neck, the beak blows and violent scents of the pigsty or the chicken coop; thus social hierarchy preserves, invariable, the trace of animal dominations and reproduces them in their oppressive course. Entering into palaces, whether royal or presidential, nay, into administration or business offices, nay, into my dean's or the president's place, I recognize there, immediately and without any mistake, the chicken coop or the pigsty, the hate-filled stench of competition, the dominance sequence of baboons and chimpanzees, whose social queue forces everyone to respire, with rancour or delight, the ass of his predecessor and to make his successor sniff his own. Language well says that the word 'society' derives from series or sequence. Everything that precedes and follows applies as well to the *Romance of Reynard the Fox*. The moral of each fable repeats singular customs and local mores, of course, but the lesson of all of them says 'alas, in these societies, always plunged in muck and appearance, Humanity has not yet been born'. This was the despair of La Fontaine as it is for me and dictated to him not to write anything beyond the *Fables* but about Love, *Tales* or 'Psyche': here, humanity finally appears. Choose: either Love or Bestiality; I don't know of any way out of this bifurcation.

Yet humans are in the process of being born and precisely from this. For, along with his predecessors, La Fontaine composed five types of fable: fables that assemble savage animals ('The Cicada and the Ant', with the latter putting the former in danger of death); fables mixing savage and domestic animals ('The Wolf and the Dog', in which, each of them in the same danger, the predator wants to change into a parasite and the parasite into a symbiont, the fable showing the very moment of transition; 'The Heifer, the Goat and the Sheep, in Company with the Lion', in which farm herbivores seem to be hunting a savage animal; 'The Animals Sick with the Plague', which sacrifices a donkey, a domestic animal); fables that place humans together with savage animals (the Bear smashes the garden lover's head); fables in which it is known what the humans cohabitating with the domestic animals do to them; lastly, fables in which men remain among themselves and treat each other with violence or amiability (hence 'The Arbiter, the Hospitaller and the Hermit'). These five kinds form a gradation whose scale describes the slow and violent accession to the human. Has it ever been remarked that the first of these *Fables*, savagery against savagery, opens the whole of the collection and that the last one, eremitic and generous, closes it, as though we were departing from signals of murder in order to construct, little by little, not only a culture, not only a knowledge, but finally a wisdom? Humanity delivers itself, free and alone, from the common living adventure, from this sociobiology that indeed lags a bit behind the *Fables*. Or if humanity doesn't live solitary like a hermit, it becomes a host and not a parasite, and a specialist in law, that is to say, in contracts.

Mountain lion and sea lion

I don't think anything changes when, leaving the farm, we descend this scale by departing to go hunting. Approach tactics and beating strategies would fail miserably without the refined bonds humans and animals have established with each other since the dawn of time: Gaston Fébus or Xenophon repeated, in our languages, ancestral gestures glorified by the Amerindians or the Australian Aborigines.[2] A hunter knows his game as well as this game knows its predator, so much so that together they form collectives linked by signals of intimidation, camouflage and incredible ostentation. Hunting societies don't only group together humans, but also unite them with the hunted: this is why Actaeon becomes a stag. Fables draw from this metamorphosis and this totemism, witnesses of the age of hunting, since they flow from the trough, starting from domestication. House and forest, with porous borders, communicate, as though the woods reduced to suburbs [*banlieues*], to banishment sites [*lieux de ban*] for the house, the farm or the city.

I have no experience with the art of hunting, but I have encountered, as I happened along during a walk in the wide open, wild animals: my sole experience, humble, of the jungle, my exit, rare these days, from the permanent spheres of domestication. Just as the model of the city extends across all of space to the point of devouring the countryside, so our relations with animals have brought about in them a universal fear before the manners of the Exterminator. Before this fairly recent age, a few sailors from the eighteenth century still testified that truly wild animals, without any experience of the universal Killer, offered themselves up to him without fear. The birds of New Zealand died like this, fairly recently, from settling on the fists of their enemies, whose cruelty without any risk condemned the most magnificent orchestra in creation to silence. Consequently, instead of this divide, up till recently pronounced, between domesticated animals, on the one side, and wild animals, on the other, a fuzzy and continuous set exists today, one in which the globally humanized environment leads to the second moment of domestication, in which every living thing acts as though it knew itself to be in danger of eradication.

This is why, without being possessed of an even moderate courage, I wasn't afraid and didn't tremble, contrary to what one reads in the books in which philosophers describe the emotions by means of abstract examples. Therefore, a long face-to-maw with a mountain lion, of the cougar type, forced us both, on a fine Californian spring afternoon, to examine our respective bodies scrupulously. This ecstasy made us enter into a kind of common tunnel that imprisoned us, the one by the other, with its silence and from which the world around us disappeared; this absence of any way out can arouse terror. Visibly, everything is decided, once again, by perceptual appearances. Which of the two will lower his head first, will lose his arrogance, will accept, through his flight, to take on the role of the hunted; which, submissive, will turn his eyes away; which then, by not showing his back, will win out in the display or the show …? This is what I have called the cosmetic: the bearing of the entire body, the neck and gaze, the erect vertebral column, the motionlessness, yes, the pride, nobility, slowness, courage, perhaps, if by this is understood a vague sum of these perceptual impressions in which size, colour, the imposing presence of the coat, the calm of the face and of the facing become all-important and all-meaningful in the vicinity of death, a thing that's still entirely corporal. When, slowly, I lifted my arms while raising myself up on to the tips of my toes to add a few cubits to my height, the dangerous animal moved away, slowly as well, so as not to suddenly lose face either. One doesn't fight with such a giant, it seemed to regret. We left one another after having first perfectly shown and then examined our bodies, honestly, without any disguise or camouflage. Like the Wolf and the Dog, the one, with bones visible under its skin, returned to its forest freedom, and the other, a slave with a hairless neck, went back to the city. We left one another after having gotten to know each other.

Prepared to know [*Parés à connaître*]

Therefore knowledge begins with 'display' [*parade*]. Delightful and irksome, this word descends from society to the court of love, then to simple postures of the body: show and vanity, exhibition, parades and ornaments, ceremony, review, chatting up and seduction, defence and counterattack, sign and pantomime – so much for concrete appearances [*apparences*]. But the meaning of the word broadens to its semantic family, in which 'apparel' is found: my yellow shirt and its ocellated coat; 'to set sail' [*appareiller*]: in which opposed directions did we move off?; 'imperial': what savage dignity did your bearing show?; 'partner': this brother-animal before me, in a private conversation apart from others [*en aparté*], precisely; 'participation': so what were we doing together, in defying each other, if not entering into reciprocal subduing?; 'parent': what symmetrical twinship linked us?; 'parturition': what new bodies did we give birth to after such an encounter [*une rencontre pareille*], both of us pregnant with a new knowledge?; plus the verbs 'to separate', as we did, alas, and yet to our mutual relief, or 'to prepare' … More than any other, this latter act keeps me, for in these corporal, gestural, postural displays [*parades*], a culture and a knowledge are in fact being prepared, begun, started, and thereby emerge. In sum, the word, all by itself, recounts, better than I do, my little story and draws a philosophy from it, as though the tradition already knew them.

In total, the only thing that got me out of this difficult situation was my old farmyard culture, all suffused with displays, roosters, boars, geldings and billy goats. So my former analysis, merely sacrificial, of bullfighting was lacking the *traje de luces* [suit of light] and the animal's black coat, the shape of its horns and of the muleta, the noble attitude of the matador's arched lower back and the raised taurine head. Everything here, once again, is decided by display. Don't reduce it to the sexual dances that instigate copulation, like the quivering red neck of male frigate birds, for, in this way, we would reduce the animal to the physiology of reproduction and to selection; no, this amorous seduction is also an integral part of the calls, signals, feints and scents whose varied rituals construct animal cultures, which are so contiguous to human cultures that they easily connect to them, on the farm, in hunting and through risky encounters. Again and lastly, doesn't mimetism allow one species to 'understand' another and, through a subtle substitution, to make use of this 'knowledge'?

Animals teach us that knowledge is born with the appearance of the body, muscles, shape, gestures and movements; display prepares it [*la parade la prépare*]. Or: there is nothing in knowledge that isn't prepared there.

Charms

Do we have here, in addition, a true genesis of our idea of beauty, the beginnings of charm's irresistible hold over our stupefied bodies? Thus, in and under the water, I have danced, in a stretch of ocean off an island in the Galapagos Archipelago, with a sea lion, supple, seductive, fluid, enchanting. When and as she curved her body, I tried to bend my own; I initiated a swimming gesture, which she immediately imitated; I went back up to the sunny surface with a scissors kick, and undulating, she would pull me down into the blue-green of the depths, beneath the cliff of turtles, amid the schools of red and green bream. Opening up countless virtualities, the three weightless dimensions unfurled before us an equal number of new choices for other choreographic paths; quick together, slow sometimes, lightning-fast at the same time and always with the same rhythm, we heard, I think, the same music, while we both thought about love, as though close to making it, beyond the barriers of absence and species.

Apuleius recounts, and after him La Fontaine, that Psyche, as well, made love, always in the dark or at night, with a kind of invisible monster in a tale where humans metamorphose into animals. Yes, I knew that the sea lion saw that I was watching her observe me; she knew that I was swimming like her and like she wanted me to swim. What should we call these submarine looks, these signals of shapes and of motor twistings, these calls of two different bodies immersed in the ecstatic weightlessness of the same elementary volume if not the emergence of a charm before the initiation into a semaphore, the first grains of a sol-fa or the ABC of an alphabet?

When, in his *Fables*, La Fontaine sings that the Lion, the Crow or the Little Fish speak, I have believed, ever since, that he speaks truly. For, during this miraculous pas de deux, we made confessions together. And since I anticipated the sea lion's gestures and she foresaw my orientations, the way the best dancers know how to do, she understood me as much as I was able to understand her. Captivated by the immediate and reciprocal reproduction of our figures, we danced while remaining face-to-face for more than twenty minutes, a long adultery – understand by this reproved word that I became an other and that the other became me – up until a male and his three tonnes of black mass came to remind the female of her duties and the human of his specific and stiff solitude. Cowardly, we moved away from each other slowly, sad from having gotten to know each other so quickly and, to our distress, so briefly.

The other

Even though we never touch, I love to walk, in silence, in the aura of your scent; I bathe with delight, to the point of drowning, in your eyes; I quiver, beneath the meaning, from the musicality of your voice; my bones crack from feeling your long-limbed and slender body, shooting high via the halo of your wild hair, move next to me; what joy to pace along, as though in a pas de deux, the hill and the plain, the bushes and the forest, the beach and the rock, with supplely adapted strides, with the same bendings and in rhythmic breaths. Enslaved bodily by your dancing sea lion charm, your wild cougar speed, your bull power with its head raised into the light and death, you enchant and torture me before we talk about interesting abstractions.

Abstractions take on meaning by resurrecting from the elementary link whose iron ball encloses our bodies beneath an arbour [*tonnelle*] from which the world disappears and whose vault echoes with signs and laughter, with colours and gestures, with untamed violence, with scented breaths. Become an animal, chained up in this tunnel [*tunnel*], under this basement window, behind this dormer, through a savage desire, I fight with you, you who are already metamorphosed, in order to free myself, by heroic and rare bursts, from this urge and to enter, streaming with meaning, into the paradise of philosophy, whose dialogue seeks, precisely, the site of reciprocal domestication, the common path by which the odorous and wounded flesh, devoured by burns and rent by noise, tears itself away towards the shared house in which arguments become clarified and harmonized. Without this prior enchantment, what weight would reasons have? If not from animals, from where do angels arise?

The others

Even though we never touch, I like to speak to a public that's multi-headed like a Hydra. The more numerous the eyes, the more intense the fascination; the more terrifying the stage fright, the more exquisite the contact. Yes, I want to flee. Then begins the body-to-body of discourse with and against silence. Everything happens beneath meaning – or in the meanings. The matador challenges the six hundred and sixty-six bulls, calls them, gets them worked-up, makes them tremble with rage and desire, arouses their ire and their stupefaction; now they are racing down the lecture hall's slope, blind and deaf, in order to charge at the low muleta, to the left, to the right, manoletinas and mariposas linked at length, supple gestures, expertly completed, which follow the attentive paths of comprehension, up until, exhausted, the bulls withdraw, suddenly tired, become icy again, beneath the heights and into the depths. Then my body returns to the dance, challenges, works up, calls, strikes the ground with its foot, announces with its hat that it is going to kill the animal at the stated hour and prescribed point of the circle, and, again, the mob of bodies, mad with rage or enthusiasm, races like thunder down the steps of the domestication house, roosters, dogs, billy goats, cows, hens, turkey, ducks, cougars, lionesses and bulls – each individual alone being as good as a species – all of them charging down at the call of the master, tamer, farmer or torero, in order to attack, with lowered heads, the red cape and sword. No, conversely, all of them, the mob of children, males and women, rush upon the sacrificial victim, the solitary bull in the centre. Would a lecture course, a discourse, an argumentation, even meaning, exist without this first face-to-face, skin-to-skin or body-to-body, without these savage fights, these crude ostentations, this hunt, this dance, these erotic laughs, these tamed cries, these learned dodges, these conventional camouflages, these collective imitations, this sacrificial ritual …, which bring the manners of the forest and then of the farmyard to the lecture course as well as to the court?

Does the one who uses these manners love them? Maybe he hates them and only takes delight in them at the moment of fusion, against nature and full of abstractions, of the human in its suit of light, alone, with the beast made up of domestic and proud animals, the monstrous Leviathan, or, conversely, at the moment of fusion of the sacrificed animal, in the middle, with the mad crowd descending the lecture hall, and, as a sum of these two figures, at the moment of fusion of the centaur formed from the human and the animal, in the central circle, with the multiplicity of animal *sapientes*, the final product of the reciprocal domestication in the common house of culture: farmyard, tunnel of encounters, arena, temple, school, amphitheatre, orchestra, theatre, political debates … If not from these animals, from where does wisdom arise?

The dialectic of the young elk and the old elk

Hegel described these attitudes marvellously but didn't put them at their true level, for the master–slave dialectic doesn't characterize human behaviour since the struggle for recognition brings into opposition, as well, two roosters, beak to spur, two kangaroos, French kickboxing, a young elk and an old one, antler to antler against each other – and not only for the possession of all the females. So when two *sapientes* come to reproducing such a face-to-face, they are reviving in and between themselves a behaviour that's common in the animal kingdom in order to obtain a prestige, one most certainly highly prized by their respective collective but in total bestial and devoid of information.

Why? Because nothing new comes from this common display except for a provisional answer to the question asked ad nauseam: Who gets dominance? Since no one can remain always master, it goes to this one or that one according to time. The repetition of the question and of the corresponding behaviour produces nothing but the eternal return of musical chairs within the species's programme. Self-consciousness occurs in neither of the two cases, not for the animals (we would have some evidence of this), nor for the human adversaries since acquiring this prestige does nothing but repeat a more than ancient programmed schema, inscribed in baboons as well as among dingoes, in which the dominator remains as animal as the dominated animals.

Or rather, it may be a matter of trying to generate a consciousness of this type, but this method fails and only leads to the dismal resignation of slavishly executing an order and, thus, of belonging to a horde organized by a hierarchy, which, in evolution, begins very early with living things. Nothing comes from this programmed automatism.

Love, not war

The fact that these behaviours already draw towards a culture and that they form part of the genesis being followed here, this, of course, is something I grant and even demonstrate, but this shred of a path leads to a dead end: the animals remain in their ecstasy, and the humans regress to their instinctual programme; this is nonetheless of use to them for putting themselves into relation with each other and therefore for fraternizing beyond the borders of their species; for nothing could be more effective for reciprocal domestication and metamorphosing humans into animals than the master-elk and slave-stag dialectic.

The need for recognition drowns us, with all hands, in the lake of species; to emerge, on the contrary, self-consciousness takes another path, more hidden, one which presupposes prestige and hierarchy to be forgotten. The face-to-maw with the cougar reduced both of us to the reproduction of learned displays; therefore we parted as vain as we were at the start, whereas the body-to-body with the sea lion made us invent a thousand new gestures in the three-dimensional liquid cube. She taught me what I didn't know; maybe she in fact learned gestures unknown to her marine species. This presupposes charms and enchantment, therefore renouncing the site of mastery, or, an even more difficult thing, a certain sharing of this place at successively ceded times. Self-consciousness is born from this new reciprocal domestication.

For consciousness [*conscience*], in the strict sense 'knowledge of the with', becomes modified from then on. Measuring the expertise of others by the modesty of its own inadequacy, consciousness remedies this by meditating in a new mode, which brings about this temporary self-consciousness that's contemporary with the double consciousness of the other and the things dominated by it. Not launched over, seeking dominance, but on the contrary thrown under, the true subject then appears, whose breathing becomes freer from seeing the other accept, time-sharing, to throw himself under in turn. Before stealing a lead on the other, a dance cedes the lead.

Flashing on and off with and according to the other's flashing, the subject is born from the other and the other is born through it, and then they both survive fed by this vibration of blinkings and occultations. The subject knows itself as soon as it recognizes its subjection while the other also admits its own subjugatedness. Heavy and black, the master came, quite precisely, that morning, to interrupt this genesis via models, modal and commodious [*commode*],³ this slow process of acquisition in *quomodo* – how do you do it? – whose progress dies as soon as the search for mastery appears. We never learn, nay, we never utter a word of philosophy without continuously asking the question *how*, the single question of modality.⁴ For in fact there only exist modalities of consciousness, to be modulated by sharing in turn. And when prestige arrives, it arrives simultaneously with behaving like a stupid ass, a phrase that's therefore justified a thousand times over. In myths and the *Fables*, the fight for prestige brings into opposition humans who metamorphose into animals through loss of self-consciousness. How much blindness must one have not to see this obvious fact that the fight for recognition, which no one can give since everyone wants it for themselves, inevitably ends in the loss of all consciousness and in the ecstasy of a drug? This transformation into a brute forever closes off, although repetitively, the genesis of self-consciousness, which is what I wanted to prove.

Psyche, on the contrary, creates it, whose title could be translated as *Genesis of the Soul Loved by Love*. For what known relation do the preceding words, with a great deal of sparkling, describe if not love affairs and their tribulations, pain, absence and journeys? Here is the dialectic, scarcely known although powerfully effective, of Cupid and Psyche, in which no subject appears, in fact, before an other has said to her, in the night, without seeing or understanding him, that he loved her. A child is born as an object body from the womb of a woman, then, a second time, as a virtual subject through the words and gestures of his mother's tenderness, and becomes an adult, a third time, as an actual subject on condition of the words of love of his mistress – a name that's therefore justified a thousand times over – formerly a sea lion, cougar and bull, a mistress he himself brings back from these faunlike forms into an actual subject through reciprocal words of love, Beast becoming Beauty again. No one derives prestige from this; the collective ignores and laughs at it, but each of us then creates his own existence, his incomparable culture and his singular knowledge. Love–re-creation–re-creation. According to the moments of the shared flashings, the common house of Cupid and Psyche will be built on the heights, vast and packed with riches and works of art, or as a poor shack in the mountains, in wandering solitude. Self-consciousness and the consciousness of the other remain an unhappy consciousness: Are there perfect great loves? Fighting for recognition, Psyche's sisters die falling from casting themselves towards those heights.

The mystery of the death immanent to knowledge

For a common death prowls around all these houses of culture: the farmer kills hens, calves and pigs; unaware that I didn't want to hunt them, the cougar was prepared to tear me to pieces, and the male sea lion was ready to expel the intruder from his gravely offended niche; my body sacrifices itself to an upkeep that no longer experiences any break; public discourse stages an imminent lynching; each of these displays, including the dialectical one, issues a challenge to the death …, and the farmyard carries on because the farmer postpones slaughtering the year's geese and turkey hens, thus surviving a season; our knowledge of wild animals depends on the suspension and failure of our hunts; and their hunt often fails, making them the crueller for half dying of hunger; the lecture hall fulfils its function because the crowd postpones lynching the speaker, safe and sound for one more discourse; in all, culture is formed through an endless delaying of death. Does it carry on the way a species does, by sacrificing the individual? These body-to-bodies, by which knowledge comes to be, imply murder. Yes, evil and deadly violence have been linked, from their beginnings and all along their courses, to knowledge and cultures, like an original sin. The problems the sciences are encountering today, from the atomic bomb to the eradication of species, recall this primary trace. This original sin always takes place.

Since two major principles expressing constants of force exist in the sciences, do we have an invariant quantity of violence? Perhaps. In the mechanical or thermodynamic sciences, these constants ensure the rational treatment of any problem by permitting equilibriums and equations. This invariance of the quantity of evil across collective spaces and historical times would likewise ensure the rational treatment of social questions. Evil permits reason, which in return refers to the constancy of evil. Therefore reason remains perverse as long as evil is maintained. To break this necessity, we must break its invariance, therefore overturn both the rational equilibriums and the equations of vengeance; we must invent the deviation of an unstable leaning, which will give birth to another state. The first person who turns the other cheek to the person who just struck the first cheek destroys the equilibrium of accuracy [*justesse*] and of justice and starts to dissolve the constancy through this heroic attempt. As an added benefit, who here becomes subject if not the person who throws himself under? Abandon prestige and domination, and a new reason will be born. Does it still remain plunged, drowned, sunk in the lake of species? Of course. The campus, the laboratory, the library …, these are houses of culture similar to the clearings where stags gore each other or to the farmyards where pigs squelch about, drugged with mastery just like the elites. We only free ourselves, anadyomene, from this archaically programmed drowning by abandoning to the animals the repetitive fight for recognition, the eternal return of prestige, which endlessly returns to animality. Even in knowledge, the work of the negative dies from repetition. Even humanity as a species can only be born from Love.

Loathing and delights: Where do we come from?

He who has looked after cows and lived in the countryside, yet without loving hunting, has therefore practised the societies, life and cultures of animals; he can, at leisure, talk to birds, like Saint Francis, ask peacocks to unfurl their fanned tails, stand at attention in front of a cougar or dance with a sea lion, in all laugh at animal-machines, while reserving the right to rethink this machinery. A peasant, sailor and Franciscan, I love animals with all my body. However, a fear of dogs, imitating with their teeth the passions of their masters, makes me hate pets, and I read, conversely then, and with ten delights, Claude Bernard and biochemistry. A scientist and country mouse, I loathe animals. I don't know which I abhor more, their domestic slavery or the misfortune without pardon of their skeletal savagery; I hate in humans what brings them closer to this abominable state.

These living beings, our fellows, arouse pity because they never leave fascination. Do they enchant us because they live enchanted? In the depths of the weary eyes of cows, dolphins and otaries, in the crazed gaze of the old male gorilla with his whitened hair like my own, with whom I spent a long afternoon on both sides of his bars, I recognize an intoxication that weighs on their existence and plunges it into a waking dream, as in an ecstasy. Do they truly see us when they look at us through that narrow skylight, through that dark basement window cut into a high and low wall that crushes their mouths and despairing foreheads? We survive from hope, whether short or eternal; we evolve with the lightning-fast rhythm of our thoughts. The only hope they can conceive is over the millions of years of evolution that leave unforeseen a few rare and contingent changes. Their interminable time leaves no place for hope. What a nightmare it must be to live like that, as though strongly drugged, relentlessly forced to eat, kill, fight to survive, sacrificed to chastity or the exhausting ownership of all the females, therefore to dismal reproduction, to the implacable law of the living jungle! From this low yard, from this cage-cellar, from this prison where the coevolutionary links between bodies form sovereign, rational and impossible to break chains, we freed ourselves one day; by what miracle? What unimaginable ancestors left, one day, from this dungeon bottom, and how did they discover, suddenly, beyond what desert, what Promised Land where the chains fell off before incomparable riches of milk and honey? Must we think that, on that paradisiacal day on which we were reprogrammed, we became a species of wanderers ceaselessly in quest of the same liberation, by another window, again? We leave the doghouse because the tension of the same chains is repeated there. We invent because we transformed into this nomad animal that departed for discovery, the first invention. No more clearing, no more farmyard, no more house, no more village, the road, places without any trace of paths ...

On this side of the temporal thickness separating us from this deviation, from that unstable leaning favourable for departure, I nevertheless think I remember this ancient jail and the terrifying archaic moments when, thrown into the animal pit of the fight to the death for dominance, amid the musky fragrances of the males in rut and the females quivering before the braids, crests and feathers,..., I risked living like this, with the same design as so many of my fellows. What thaumaturge freed me from this? For what aimless journey? What Love raises *Psyche* or the Soul out of the deadly lunacy implied by the bestial jealousy of her sisters? What breath of zephyr, in lifting her, deviates the girl away from the tragic gravity of the *Fables* in which her family lives? She throws herself from the cliff; the wind gathers her up again and lifts her: without any miracle, she deviates from her equilibrium. When her sisters try it one time, they break their necks at the bottom of the precipice. Proof of the risk run in the game of such a deviation. What contingent event made us leave, likewise, the rational reduction to inflexible mechanism, the laws of the fight for life, the pressure of the better or the best adapted, the necessity to have the most children possible in order to win? What deviance from animal laws made us into exo-Darwinians?

Did liberation come from the genetic path of signs, which, by rare bursts, animals took as well and which, sometimes, can lead us all the way back to them because, precisely, we followed this path all the way up to the risky moment when we left them? Carrying colours, gestures and scents, this route traverses the basement window of their eyes, the orifices of their sense of smell or of their heat-sensitive organs, and passes through the light of these narrow skylights; a few calls, sounds, certain words also cross their hearing. What chain prevents them from pursuing this route? What drug obstructs them from pursuing its direction? And yet the robin has an enormous repertory: more than a thousand different songs; the marsh warbler and its neighbours organize polyphonies that arouse even different species of winged creature. Groups of chimpanzees hoot and pant at night in choirs that form and come undone. Universal among humans, like mathematics, music therefore extends at least to apes and birds. How, endowed with such a small brain, do these latter attain this complexity of codings? Because their bodies fly. Thus they live height, descent and ascent, vibration and holding, the rhythm that defends its regularity against the turbulence of the flows, the mobile equilibrium of a glider, a freedom in space imitated at leisure by the freedom of modes and tones. In the three dimensions otaries dance and birds compose; but apes, as well, fly from branch to branch. Why, singing in this way, didn't they fly away all the way to us?

Hatred, love, survival

A shy animal, barely cured from animality, disintoxicated, unchained, I remain, although saved by the deviation, an animal still. Since life, in deviation from matter, nevertheless follows its laws, my tilting outside animality leads me back to it through some inevitable gravity. I can transform myself into one of them by means of that enchantment that causes a person to lose what philosophers, with arrogance and anxiety, or perhaps scorn, call the 'self'. So I can communicate with them but quickly grow weary of it. I feel a passion for, but above all I feel a deep fear of this metamorphosis and being plunged into this dialogue. How can one love explosive life to this extent and turn away from it all the same? Because it only exploded in order to obstruct itself. I participate in this animal culture; I know it, love it, am fascinated by it, remain nailed to it. But I am terrified of this loss; I have always wanted to turn away from it in order to discover, by forgetting this cruel origin, another knowledge, new loves, the absence of suspicion, a happy contingency, potential multiplicities. This freedom of polyvalent signs liberates from the single paths, rational and inflexible, of instinctive programmes. The light mind of the virtual turns me into a sur-vivor.[5]

We certainly love this first, originary, conditional life but detest its rules. I notice this ambiguous feeling towards life everywhere. In science: we know its constraints but no longer interrogate it in our laboratories, devoted to mechanism, as though nothing was going on. In politics: the implacable laws of the strongest, of the best adapted, of the winner, of the one who multiplies in order to invade time and space, the Olympic Games and their medals, states and their power, capitalist businesses and their triumphant size fascinate for a moment, but disgust forever; the number of people killed by Alexander, Caesar or Napoleon, those bedecked roosters, the deportation to Australia of convicts treated as though they were at a slaughterhouse, the gulag, the Shoah, Hiroshima, this bloody history resembles far too much the life of the worst of all possible animals, the one that abandons most of its children to death. We love life, we say, but don't we abhor that particular life, nevertheless the same and following its laws? This is why, lucidly, the most distinguished of our wisdoms seek to free us from it. Through the same ambiguity again, we claim to cherish animals, but through individual sacrifices and extinctions of species, *Homo terminator* destroys them and ceases to converse with them, excludes them from its cities and habitats. Thus, lastly, this living body, our beloved hotel, half-in chains half-free, we hold it dear and we hate it, suicidal in the majority of our actions. In our bosom dwells the hell of the *Fables* and the heaven of *Psyche*, a Plato buried by the body like a grave and Christ himself, God incarnate, emerging from the tomb with a glorious and immortal body. We seek to resurrect from the death implied by life: we hate, not life, but that death it drags along with it, the one we undergo, the one we can inflict. We try to escape these deaths, which acts of parasitism, dangerous encounters and sacrificial gatherings drag after themselves.

Human knowledge and culture emerge from this ambiguity, are therefore born from this deviation from animal culture and animal knowledge, often disappearing into it by collapsing back into it but incessantly being reborn from it; this emergence dates from time immemorial but always takes place, this morning as well as long ago, humans having difficulty in being born and therefore having to be reborn often, from hominescences to hominescences. We live in the immemorial night of the *Fables* for three-quarters of our actions and attain heaven so as to immediately leave it. The angle of the deviation allowing this difficult ungluing is measured in attraction and repugnance, sleep and anxiety, bestial life and sur-vival, two ecstasies: in Hatred and Love.

Contingency

Nothing could be more logical, rational, rigorous or necessary than equilibrium since it ends the path of falling bodies. By their etymology, both 'ecstasy' and 'existence' designate a deviation from equilibrium; they therefore free themselves from necessity. How? Why? Will we ever know, since this deviation defines reason itself? In principle, reason in effect asks why something exists rather than nothing, without suspecting that this question contains the answer: 'to exist' as well as 'rather', both express this deviation. And since this deviation appears, like Lucretius's *clinamen*, *incerto tempore incertisque locis*, at uncertain times and places, it carries contingency along with it: it could well not appear.

Earlier I mentioned the greatest contemporary event. To finish, I would like to express the philosophy of the Event itself. Deviation from equilibrium constitutes the contingent event that breaks with necessity, the necessity of equilibrium, to be exact. Now, contingent means tangent to the necessary, touching with its own curve the legal, rational and repetitive straight line at one point and – deviating more and more away from the point, the straight line and their necessity – departing in this way, where its fancy pleases, amid the possibles. Thus our physics and biology try to explain a world that's so contingent that it can only give itself up through precise experiments. Thus the bushy arborescences of existence and knowing run from contingency to contingency, world, life, species and cultures, knots where each bough deviates from the trunk and where each branch bifurcates from each bough. This book explores one twig of it.

Two ecstasies, two existences, that of animals and our own, are therefore distinguished, literally, as well as two deviations, without which neither life nor mind would appear. Two survivals likewise differ; someone who has survived a typhoon at sea, a thunderstorm in the mountains, an earthquake, a despair over a failed love lives close to nothingness, in the vicinity of the final equilibrium, so that surviving for him repeats with difficulty, at the beginning and starting from a few pieces of retained information, the deviation from which all life appears – wild, basic, elementary – and even all world appeared to the light; after having escaped death, life reties with life [*vie*] the way an overcast stitch [*surjet*] straddles two fabric edges and sews together two selvages. But, in order to justify its prefix from *survol* [flying over] better, survival [*survie*] can, in turn, make another deviation, a new leap, a strange jump. Through this jet that shoots up and creates, shy, fragile, yet powerful, survival no longer escapes from death, but from life itself and, as a result, from the death implied by life. So, survival leaves life, abandons it, remains, of course, subject to chemical and vital laws, but overhangs [*surplombe*] them, overtakes them, surmounts and flies over them, supernatural [*surnaturelle*]. The reign of the mind begins.

Modalities

Mechanical laws are universal and necessary: gravitation and the second law necessarily bring the existent back to nothingness. For inert things to exist, they must contravene these laws, which, in the square of modality, only occupy necessity's corner. So new physico-chemical laws, those of existing things, possible and contingent, come into play with or against necessity, in the other categories of modality. Contingent therefore, the laws of physics and chemistry, not necessary but universal, occupy the totality of the existent, all of space and all of time, occupy forces, in brief, the set of known units. For living beings to exist next, they must deviate from physics and chemistry but while remaining compatible with them; they therefore obey the same laws, but the laws of living beings, as contingent as the laws of the inert, not universal like them but merely local, require the closure of a system that's partly open, like a house; this is a second state or stage of contingency, a kind of second differential. The free play of the laws invades all of the possible in the inert, consequently this free play is universal, whereas it is circumscribed within the circumstance of a place and a time for living beings. Life brings into play physico-chemical laws within the constraints, local and temporal, of a defined system, closed because living, open because compatible with said laws. The order of the living remains physico-chemical, but the specific originality of its secret is held in this circumscription of place and time. The modalities will have to be rethought.

II. Transition

We no longer recognize the animal races

To nourish a growing population, artificial feeding in indoor breeding farms of selected animals began starting from the nineteenth century in Darwin's England; in 1854, an essay by the French traveller L. de Lavergne[6] describes them and relates that they aroused there the same opposition heard today regarding the same subject: pity for the beasts, the bad flavour of the meat so produced, the risks of disease. But frequent epizootics caused the failure of these attempts so that the generalized revolution of livestock farming dates, as had to be expected, from the Second World War.

Today, what we call industrial agri-food, a name that's too general to cover every situation, has definitively taken the place of the domestication I just described. Even before we modified the genome of living things, the zootechnicians themselves, from the 1980s onwards, no longer knew how to recognize hens, pigs, turkeys or cows. Shaped by the inspired cross-breedings of population genetics, refined by productivity, displayed in advertising, asepticized by precaution, crushed by administration, domestic animals seem, from this period onwards, to be as changed as their masters. I remember the date when I had to run up and down the entire department to find a hen that still knew how to brood after laying her eggs.

Animals changed bodies just like we did. And we changed worlds together.

Inhabiting: The end of codomestication?

Above all the very process of domestication is changing, the domestication of humanity and of its former companions; in brief, the house is transforming. What does the verb 'to inhabit' mean? We haunt two niches: our body, whose skin and fragile loves we shelter under a roof, between four walls. Inside this construction, drawing the boundaries of energy and space–time, lived every species, bathing in the same warmth and pumping into the same sources of power across a closed and porous thermodynamic enclosure. But this vital complex of energy and information has disappeared over the course of the last half century. Travelling through Queyras around the 1950s, Robert Doisneau didn't suspect he was photographing the last habitats in which cows were being used as heating for humans.

Excluded, from the inside to the outside, from the mother house, chased away, from the outside to the inside, from the traditional pastures, animals live among themselves, like we do. They are in boarding school, while awaiting the slaughterhouse, like us and our young, schoolchildren in middle school, the sick in a hospital, adults in an office or workshop, the aged in an old folks' home. We no longer live with animals. No one lives with the other any longer; we sleep among our own group. We survive under a system of separation. We share with our former companions the consequences of this sudden or slow rupture of codomestication. Therefore, when zootechnicians no longer recognized animal races or individuals, doctors no longer recognized our former bodies, soon drunk with immortality. The last days of the first domestication, the first beginnings of a second one: begun thousands of years ago, the one is coming to an end; the other is beginning. Different bodies no longer inhabit as their predecessors did.

Provenance

For our bodies lastly inhabit the world. The farm marks a precise stage in the contingent process of hominescence. An example: lying there, living there, we knew the provenance of our food, stemming from our garden or the neighbour's; we knew where the wheat in the bread came from, the milk in the cheese and the medicinal plants used in the event of illness. Domesticity then designates a set of living things which depend on us and on which we depend. Inhabiting consists in knowing the provenance: of sons-in-law and daughters-in-law, of species, of course, even the provenance of ideas. Sugar, resin and rubber flow, sticky, from the barks of maples, pines and heveas; we milk milk from cows that have just given birth, steal honey from beehives, extract our equations from experiments. Provenance guarantees worth and not only of wine.

We now eat food produced on the other side of the planet, whose origin, growth and flowering are unknown to us; our meats and fish come from distant boarding schools, our vegetables from complementary seasons and our fruits from exotic greenhouses; we eat pears from the other hemisphere the way our bulls eat a soybean meal grown far from here; we treat our diseases by means of medicines studied in laboratories scattered across the planet ... In losing the finitude implied by the definition of place, we have left provenance. Our sources of survival expand into the world, like the sources of our crops and our animal husbandry. Thus the parallel of the three organisms, animal, floral and human, is coming to a close, living, in fact, always together, although apparently separated.

So we all moved from the 'there' to the world; better still, from being to modes and relation. In a way, neither humans nor the species have left the house, I mean that they and we are again inhabiting the same space, except global, an open thermodynamic enclosure whose laws are unknown to us and gods we don't see, but from which we draw companions, food, medicine and reproduction. We used to inhabit the peasant house; today, the former living things of the household sail like seafarers, between walls whose wavering causes the adaptive body of the sailor to settle in and travel at the same time. Life used to inhabit Being [*l'Être*] archaically; it is evolving and wandering in Relation. It used to haunt the layout [*aîtres*] and the hearth [*l'âtre*] of the farm, which was cut off from all ties except close ones. Did we know that the loftiest ontology was still tied to humble peasant practices?[7]

Again, this transition can seem slow or abrupt: the English language reputes the fowl French says is from India to be from Turkey, and Spanish says it's from Peru, and who, without ever having seen a dog, would believe the Dalmatian, the Newfoundland, the Poitevin, the Shetland Sheepdog, the Great Dane, the Labrador or the Havanese to be of the same species? The former globalization, a movement of intercontinental exchange of fruits and vegetables, of spices, tea or coffee, suddenly became generalized a few decades ago in order constitute a world conspiring with new bodies.

The fragility of the 'good plants'

Likewise, the dispute, so recent, over the new bodies called GMOs awakens memories lost ever since the Neolithic. Some given variety resistant to the 'bad' plants, weeds, or enriched with vitamins is prepared on the bench of a laboratory, whose enclosure and white coats recall analogous behaviour lost in the mists of forgetfulness.

The new organisms of the dawning agriculture, corn or wheat, discovered or invented I know not how, must have stood out for their fragility. If not, why delimit a field, plough it to exterminate the species that could have supplanted their rising? Why look after steer or pigs inside an enclosed farm, well named?[8] The first farmers and the original livestock breeders took and handed us down so many precautions because this 'domestication' brought about varieties that were so frail they had to be protected so they wouldn't die the moment they were born. Toss a few grains of wheat randomly in various ditches, even in Beauce,[9] or let this heifer or that piglet run loose in Africa in the neighbouring forests – how long will they survive? A quarter of an hour for the animals, a season for the grains. Agriculture and livestock farming were born of the fragility of the domesticables. Noah's Ark, into which the patriarch crammed the species, symbolizes the barns, granaries, sheepfolds, stables, enclosed fields, patches of alfalfa, walled gardens, in which *Homo rusticus* protected a few living things from the diluvian furies of reciprocal hunting, called the law of the jungle. This particular law devours little ones the way other laws graze their wheat on the blade. The first person who, having enclosed a piece of land, took it into his head to say 'this is mine' invented the mastery of the domestic living thing by defending a precarious and vulnerable species. I wouldn't swear that property didn't begin at the same time as this protection did.

One more time, like a thousand other times, these changes come from weakness. But why the devil did these species become frailer than the others so quickly? Precisely because we protected them. Because, just as we ourselves had done, they had left the laws of nature such as Darwin later discovered them. Agriculture was born then from a loop brought about by precariousness. The livestock breeder and the farmer protect these species owing to a vulnerability that increases intensely with protection. The hominization of *Homo sapiens* must have followed such a circle. And, once again, why this weakness? Because these species, like the human species, increasingly quit nature as the protection became increased. Why, lastly, this weakness? Because these species, like our own, are artificial; yes, call them technological. Ever since our dawn, we have eaten a technological wheat. As artificial as oxen and wheat, even more precarious and vulnerable,

GMOs haven't benefitted from thousands, even millions of years to attain the adaptive power of selection. Domestication in a way breaks with the law of the fittest. The enclosed field and the walls of the barn protect the least fit, which become all the more disadapted the more these enclosures guard them. Thus, humanity can be thought of as the animal who builds a house: protected behind its walls and under its roof, its weakness gently leaves evolution; our domestication precedes the domestication of animals, and this latter domestication consists in having chosen plants and animals, those first artifices, enter into our house or into the ark for the purposes of protection. But who, holding forth on technology, willingly cites corn or oxen?

Today we are rediscovering the same mastery of the living thing; this mastery worries city people because, only knowing concrete and glass, they imagine technology to be foreign to the living. We used to have command over selection and in the long term; now we intervene in mutation and in the immediate term. This variation changes the rhythm and scope of our actions, not the actions. The protections and the artificial, the weakness of the least fit remain constants: the enclosures of laboratories replace the enclosures of fields and of pigsties; the researcher's white coat replaces the farmer's boots; high-precision equipment replaces ploughing implements, and the extreme fragility of genetically modified organisms replaces the congenital vulnerability of phenotypically modified organisms.

Are you afraid of what we have to protect? Were you afraid of oxen or wheat?

III. The new totipotent culture

The new common house: The world

So we left the farm and entered the world. Again, how? This is how.

The enormous background noise of shrimp and krill covers over the murmurs of the sea; flies don't see like us with their compound eyes; sensitive to polarized light, bees perceive ultraviolet radiation, while infrared excites the eyes of certain reptiles; the pitviper's thermosensitive organ captures tiny bits of heat; bats guide themselves by ultrasound; torpedoes discharge their electric organ on prey that subsequently become afflicted, as our Greek ancestors said, with a narcotic torpor, from which the name Narcissus was derived; the magnetic field, it seems, directs the flights of migratory birds; how many living things react sexually to minute densities? The auditory bulla and cervical spaces of whales and dolphins function in a way so superior to our sonars that they perceive, by piecing it together with the help of bathythermographic elements, a three-dimensional map of the oceans they are swimming, communicating and mating in; down in the oceanic abysses, the fights to the death between sperm whales and giant squids pit the keen hearing of the former against the gigantic eye of the latter, capable of seeing in the darkness of great depths …

I'm not trying here to say how these organisms are constituted, but first and foremost how they function for a precise activity, the activity of communicating with each other and with the outside world. Described too quickly in this way, each of their performances cuts out a respective niche in the world, a niche defined for each of these organisms and into which the organism slips, adapts, survives, multiplies, haunts a space and invents its time there, whether an ephemeral or decennial one. Flies don't 'see' what bats 'hear' nor what pitvipers smell; the life duration of whales and the length of gestation of their females diverge from those of krill – can we conceive of the time of chrysalids?

The sum of the worlds and times of every living being?

What should we say about our body? That it now receives ultrasounds or infrared, electromagnetic waves and thermal agitation. Bathythermography is part of the basic instruction of the submariner, and the spectral range of the invisible and the inaudible unfolds before the senses of astrophysicists. Our submarine listening detects, for example, calculates, hears and negotiates the enormous background noise of shrimp and krill as well as the messages exchanged by whales between each other and dolphins between each other. We intercept the murmuring of anthills.

The dividing up of the global environment by the species, from which each of them cuts out its world and the rhythm of its time, doesn't only concern sensibility but ten other behaviours: kinesin 'walks' in cells; under the effects of light, chlorophyll synthesizes carbohydrates from carbon dioxide; proteins adapt to each other via stereospecificity …, all of them corporal performances breaking with the scale of our own performances but which we now know how to recognize, explore, imitate, reproduce, even improve; when our machines broadcast, receive and analyse signals in frequencies or intensities outside the range of our senses, do we think that other living things sometimes share them or can intercept their ranges?

So given the set of signals of all types that are accessible as signs by the set of living things, our various apparatuses tend to reconstruct this set as a sum of the habitats each species – including ours or each individual of ours

– cuts out in its environment. Are we thus tending, at least asymptotically, towards a global reality, towards an integral of these fragmentary spaces and times (niches and durations of every living thing) and, through having collected them together, towards the beginnings of an integration?

This is how we entered the world.

The site of a new aesthetic: The Biosom

The technologies and sciences therefore gather together these immense masses of data today as though these data came, not from our body (as the old empiricism said, when knowledge arose from its sensations), but from a kind of global body formed from the evolving sum of species and kingdoms. Through technologies, which I've said set sail from our bodies, we broaden our reception to the entire possible empirical biocapacity. The old empiricism finds itself changed from this: it surpasses the five or seven senses of individual bodies in order to extend its span to all living species. I call this virtual body, active in knowledge without us really being aware of it, the Biosom.

We are attaining here a global concept not only having the exactitude of the sciences called experimental but while respecting all the scales of a possible map of the world, a map punctuated by the spatio-temporal niches of the species. Don't you find this body in formation to be more real than the one we call real merely because we feel it to be our own? And more objective, to boot? Whether traditional or contemporary, do the empiricisms take account of the fact that we no longer have confidence in our own bodies because we transfer this confidence to this Biosom? Don't you find it to be more concrete than the body of the imaginary large animal the tradition gave to the collective? Don't you find it to be not only new, of course, but also very ancient since the sciences carry out in it the old programme of the *Fables*, which knew how to metamorphose us into every species? We are beginning to decipher the language, time and world of dolphins and of bees, of the Oak and of the Cicada, of the Fox and of the Ant … Through this kind of sum or marquetry under construction, on the 'subject' side, we tend to construct piece by piece, discipline by discipline, on the object side, the fables, our new common house, which no longer merely gathers together certain chosen species, fragile and domesticable, but the entirety of living beings.

This Biosom attempts to 'synchronize' the times, highly multiple, of fish and of bacteria, of whales and of mayflies, of the rose and of the sequoia, of birth and of metamorphoses, of the nymph, of the chrysalis and of the imago, of reproduction and of death, of entropy and of rebirth. If each species cuts out a space–time, that is, a niche, in the world and survives there, if conversely the spatio-temporal dimensions of each niche make one or several species appear, as though they were the hallmark of each step or each level of each scale, then the Biosom composes the global, complex and criss-crossed space–time of the entirety of this world's living beings. How are we to describe the observers and the observed of today if not as such sums in the process of becoming?

Intelligence and the world

To this Biosom, which could also be called Macrobe, to say the largest of lives, let's dare to add a Hylosom, a body that also sums up the elements of matter, in that our apparatuses now know how to measure the nanotime during which certain subatomic particles appear and disappear as well as the gigatime separating us from the big bang and the big crunch, the presumed birth, upstream, then, downstream, the supposed death of the Universe. We don't only enter into the niches and times of ephemeral microbes and ancient sequoias, but also into the sizes and durations of atoms and of the World in its entirety. To the temporal overlapping of the observers' bodies are added, through a thousand different pieces of equipment, the entirety of the temporal overlappings of the matter forming them. By manipulating these apparatuses, we incorporate, in a way, the entirety of their performances. By observing what the other observers are observing, we gradually complete the observed as our construction of this sum progresses. Revisited according to the real functionings of these approaches, empiricism, it seems to me, turns less towards subjective relativism or universal flux than towards an integral or a composition which runs asymptotically towards a global object, the world, even more objective than any object known to date since observed by every possible observer.

Do the existence and activity of this body undergoing integration, a body receiving a torrent of landscapes submerging our data banks, now require us to invent a new understanding commensurate with this ever-newly invented body? Under pain of drowning in this tsunami of data, must we attach a soul to this Biosom, which is constructed with patience and technologies, level by level, a soul we don't yet know anything about? The subject of thought becomes generalized then to every possible living thing, like that of a bio-cogito, but also to the Universe, like that of a cosmo-cogito. I shall soon say how the new information technologies favour new 'faculties'. But it suddenly occurs to me, doesn't artificial intelligence already carry out this programme and this function? Let's not think of artificial intelligence with reference to the brain alone but to the body, nor only to the human brain but to this Biosom or Macrobe.

Levels and maps: Riemann surfaces

The passage to the global happens here by not missing any of the intermediate steps along the arborescence or rather the flaky pastry foliation of the living and the inert. For we readily represent, at least ever since Porphyry, the whole of living things by means of an arborescence, a model that has the advantage of suggesting their genealogy but the drawback of projecting kingdoms, genuses and families onto a single one of the living things.

For why would the vital dynamic, in general, slip into a particular form of vegetation? Why, consequently, not also represent living things by means of a Riemann surface, a kind of multi-layered flaky pastry mille-feuille or pile of cards at different levels, connected transversally together? This model would have the threefold benefit of showing the global integration in formation, plus the biotopes and biocenoses running across the levels, but also the genealogy along the connections.

Historical reprise: On observation

I shall return to the observed. Whether the Sun revolves around the Earth or the Earth around the Sun, both theories, unproven, save the appearances, as the Greek astronomers used to say, who called this indifference 'the equivalence of hypotheses'. Reversing Ptolemy's geocentricism, Copernicus, without any more proof, decided to place the Sun at the centre of the heavenly motions. Galileo was the first to meditate on these decisions by conceiving a class of equivalent observers such that the appearances ascribed to the others by the reason of each of them would be identical to one's own. Whether it's a question of classical astronomy, rational mechanics, quantum mechanics or relativity, the living or sensing observer always reduces to an eye on the move, therefore to kinesthesis, that seventh sense that is movement, well known and well described today, which *The Five Senses* called 'Visit' or sight in motion. Does this observer, which it is sometimes claimed introduces life into thought, perceive colours and hues, sonorous varieties emerging from the background noise or calorific and tactile waves? This eye in motion, in a train for example and picking up a signal, substitutes for Condillac's frigid statue. Physics remains mechanistic, and knowledge goes without its biosphere. Empiricism refers to a body that's reduced, truncated, castrated, abstracted or abstract to say it all.

Starting from Locke and the Enlightenment, this formal observer, whose movements only take place in a theory of groups, enjoys senses, doesn't merely see, but tastes, touches, hears and delights in fragrances. Knowledge broadens its sources; to kinesthesis alone, this observer body adds every possible aesthetic. It's not a matter solely of speculative philosophy or cognitive science since the experimental sciences implement and fine-tune this lesson. Astrophysics and the widening of the world to the Universe emerged from the cold and warm colours of the spectrum at the beginning of the twentieth century; we owed this exit from the solar system towards galaxies to the red and blue fringes that gave away the age and temperature of the stars. So these signals of sight, and also hearing, enter into families of waves, into groups of signs. The class of observers consequently realizes that it only perceives within narrow windows and that, most often, it remains blind to the invisible and doesn't hear the inaudible, doesn't even touch the tactile … without ever asking itself the question: invisible, inaudible …, but for whom? From that time forward, this class has ceaselessly constructed receptors whose performance surpasses the class of waves that are perceptible by the class of new observers, from the ancient theodolite to the radio telescope and from the old lenses polished by Spinoza and assembled by Antonie von Leeuwenhoek to electronic microscopes, scanners or cyclotrons. Who, consequently, has remarked that this broadened sensorium covers in part the entirety of the sensoriums of the living species, that these instruments of perception can be understood, already, to be biotechnologies?

The old geometral

The observer perceives movements, as we have seen, forms and qualities. For, regarding form, seventeenth-century authors delighted in the following line of reasoning, delectable. Let's observe this vase; from here, I see it as inclined; if I come closer or move away from it, I see it as upright or oblique, and from above or below, as smashed or suspended …; you will never see it other than from your own point of view; your neighbours to the right or left perceive it differently, and distant spectators even more so. We all only perceive profiles; no one observes anything other than some silhouette, and each of us sees a different one. We can even draw the cone of vision and demonstrate, by means of geometry, that one person sees a circle, another an ellipse, a parabola, a hyperbola, etc. according to the variation of the sections of this cone; it would be impossible to tell the form of the 'real' section. Watch, even, particularly in the hard sciences, the relativity of the points of view stand out in profile. Such relativity doesn't solely affect the diversity of customs, institutions and laws, as Blaise Pascal, after Montaigne, wrote, rather the geometer no doubt inferred this from the variations of his cone sections. Far from committing the stupidity of separating the sciences from the aforesaid humanities, philosophy discovers the relativity of objective and cultural perception at the same time. The author of the 'Essay on Conics' had the truth of laws and customs vary according to the Pyrenean crest the way the circle and parabola did following the cross-section.

So can someone perceive the object as such, that is to say, the integral of its profiles? No one can. We only see figures, never the transfiguration, the sum of the figures. Only a few chosen apostles, Scripture says, saw this transfiguration, that is, a completely white body, 'whose face shines like the sun, and whose clothes become incandescent like snow', white light forming the sum of all colours. Hence, the seventeenth-century authors said, the existence of an integral observer was required, an observer who would see this form, called the geometral, or the object as such, a bit as though the apex of the cone of vision tended to infinity so as to transform it into a cylinder. They called this visionary limit, God, and this geometral, world. The divine cognitive function integrates the global whole of all the oblique observers of the same form. This God isn't quite absent, whatever may be said, – either from the sciences called hard or from the global class of observers of movement – in the formation of the ideas of space and of time. Whoever thinks time and space are absolute designates God in some way, monotheistic Newton or pantheistic Einstein for example.

I am continuing here a similar line of reasoning, generalized to a global empiricism, to every function of sensation, on the one side, and every function of life, on the other, lastly to matter itself. But instead of going to the limit of a form, I'm leading this reasoning inductively, step by step, level by level. Since we don't have a global idea of life, let's not be surprised to find it as the integral of the perceptions of every living thing. The conditions for the exercise of the experimental sciences encourage us to reflect on this integration of the sources of signals and of the capacities of reception. The class of observers slips, here, into the tree of species, into the time of matter and of life. Not only does the contemporary scientist live and exist just as much as he thinks, but his body is reconstructed in steering towards this summation, and his thought has as its object a world less splintered than it was said to be, more integrated than it was believed to be since his actual work consists in the patient construction of this white sum. Here is how.

The biocosm and the precognitive function of the Biosom

Supposing, in fact, that we 'observed' the world starting from this integral Biosom, or undergoing integration, what world would we have to do with? With the one that the living things, if they merged together, would perceive, therefore with an integral under construction of their respective *Lebenswelten*. Every living thing has its world, even if Heidegger calls them 'world-poor'; the wealth of our own comes from this integral that's undergoing summation or from the union of these poverties. The Biosom receives, in a still and always incomplete way, the signals of the world full stop. Codomestication becomes totalized.

The entirety of 'observers' therefore inhabits the evolutionary house that it observes and evolutionarily observes the very house it inhabits. So this subject on the path to globalization is heading down the path to knowing, no longer merely its farmyard, but the world itself in increasing totalization. All real and possible kingdoms taken together, life therefore realizes, by a kind of fold over the material realm, the entirety of the ways of apprehending the world or of communicating with it, already of knowing it. As though the world were slowly evolving towards self-consciousness. Does the appearance of life mark the first act of the very emergence of knowledge? Are we entering into the vague idea of a capacity for self-comprehension of the world by itself, for which we would function as mediation?

What then is life in general, I mean life such as living things in their entirety share it? The first step in the knowledge of the world. But where does life come from? From the world itself. Therefore the world produces the first link in the chain of its self-knowledge. Beginning in and by the world and perpetuated via a perception of one of its fragments, which alone permits it to survive, all life furnishes an original circle segment in the feedback cycle of the world and knowledge. We are running along the path of integrating these fragments. A question: if this self-feeding exists, does the world need knowledge in order to perpetuate itself? More than contingent, was it quasi-necessary then for the world to produce life? Doesn't this shed light, in return and in a strange way, on the double emergence of life and of knowledge? Thus the Biosom and the biocosm furnish the concepts, elementary and global, of a general theory of signals.

But, ever since our times of hominescence, the Biosom hasn't solely concerned perception.

Pangenesis

> *Homo biologicus* ... will accumulate the properties of reproducing without a male like aphids, of impregnating its female at a distance like nautilus mollusks, of changing sexes like swordtail fish, of putting out buds like the earthworm, of replacing its missing parts like the newt, of developing outside the maternal body like the kangaroo, of putting itself in a state of hibernation like the hedgehog.

To this list written in 1956 by Jean Rostand, Jacques Testart, who cites him, adds that in imitation of sexless bacteria, we will substitute reproduction by cloning for procreation, repeated duplication for chance encounters and mixtures. Both of them condemn such regressions in the scale of living beings, instances of heading downward imagined by the one and accomplished in part by the other.

There is nothing to object to about this almost universal and very rightly described pangenesis except that it repeats ancient customs. We eat grass and lettuce like elephants, gorillas, gazelles, cows and sheep, meat like lions, tigers and dogs, grains like canaries and chickens, fish like seagulls, molluscs like saurians, fruit like chimpanzees, honey like bears; the most sophisticated of gourmets devours truffles like a pig (without any offense meant for the truffles, I would never forgive myself, nor for the pig, who I respect as my semi-twin brother), and our children enjoy sugary solutions like bacteria across their membranes ..., multiparous, of course, but still omnivorous. In addition, we walk, run, jump, swim, dive, fly like ten species, less well and sometimes better than they do, and even catch up to the furthest ones at the bottoms of the oceans, in the stratosphere and the subterranean caverns on board balloons, satellites and bathyscaphes. We communicate at great distances like whales, eels or salmon. What dancer, what mime couldn't imitate the gestures of twenty animals? Sports compete against panthers and lynxes. What child has never played at Wolf and Lamb; what *Fables* do our cultures come from; what fetishes do our religions ...? In and through our acts, works and labours, postures, movements, technologies, imaginations and poetries, our potentialities open up to the scale of the living. What is there to be surprised at if the quasi-complete course of these unfolded spectrums of possibilities suddenly affects reproduction since our other functions, without exception, already bear witness to this totipotency? Yes, we are all-powerful.

Totipotency

This adjective in no way signifies that our strength prevails over the inert, existent and thinkable world, nor that we are again taking up the ancient folkloric place of being the masterwork of nature or the master of the Universe. No, here, power amounts to possibility; our capacity opens up like this. Pangenic and omnivorous, of course, but in sum omnivalent.

Taken in the sense of force, power, on the contrary, reduces this unfolded spectrum down to a single line, returning thus to the specialist niches occupied by living things and to the single order whose stupidity and cruelty we observe everywhere and always. Hierarchy, excellence, dominance, all of them linear limits, remain within the remit of the species, whether single-celled, animal or plant, each species refining a kind of jumping, running, flying or swimming in order to better adapt to some place of arid land, turbulent air or deep waters. For specialized expertise is needed to master a small niche, like a reptile, an athlete or an academic (whether chordate, runner or learned, they are easy to classify), each of them dominant in its little department. Just as animals do, divided because of this into kingdoms and genuses, a runner, an administrator or a financier projects the complexity of humans onto the simplicity of time, money or codes, the way a crocodile does onto the couple motionlessness–speed or the eagle onto the association of flight height and lightning-fast descent. Every hierarchy of this type – power, wealth or force – obeys the linear logic peculiar to animals and plants, in general to living things, all of them seeking mastery in a precise little niche. I don't know of any better definition of specialty, therefore of species, than this linear projection or this linear reduction, than this restriction to a single variable.

Abandoning this type of narrowness, humanity, I was even going to say humanism, attains capacity's multiple opening. The moment of hominescence has abruptly opened up this spectrum, from the side of corporal genesis, of the observed world and of communication. A doubling back along the usual path of living things, unpredictable no doubt, the cause and circumstances of which we don't know, had already opened us up, and quite early on, to this totipotency: should humanity move or eat, it has already risen to the general. For, from the hand, despecialized, to the entire body, from postures to movements, from eating to reproduction, from cultures to occupations, we travel across the families and orders through which life enters into species. We 'can do' almost all of the species and others as well, whether extinct or virtual. From being the sailor that you were, with your fingers in the splices, tomorrow you will become a virtuoso pianist, fingers and palms travelling arpeggios and passageworks;

but also and as well a mechanic under the belly of your car, a clawed climber, without taking into account the caresses of the lover. The entirety of our devices, themselves having set sail from our body, has the effect of augmenting, along the same line, this totipotency. How would technology dehumanize humanity since it extrapolates the best corporal characteristic from it?

Monovalent and finalized, each traditional technology therefore contributes to forming the Biosom, become totipotent through their sum. Omnivalent, on the contrary, in perfect harmony on this point with our bodies and with the Biosom, the new technologies then take over the reins. So if the old technologies, monovalent, hard and of the order of Hercules or Prometheus, slowly (I mean along a duration compatible with the duration of hominization) formed the Bio- and Hylosom, which bring us closer to the world, itself hard, the new technologies, totipotent, soft and of the order of Hermes or the Angels, quickly facilitate the formation of an Ethnosom and probably create it by bringing us closer to humans. Good news for the social sciences.

The second origin of our knowledge

Without this series of omnivalences inscribed in our bodies and its functions, and then extrapolated by various technological performances, what would we know? Adapted to its niche thanks to its organic mastery, every species, 'world-poor', follows its detailed ways, a little labyrinth cut out in a black world, whereas opening up to a brighter world is the animal that's 'capable' and white, it too, because ordinary light adds up every colour without letting them be seen. Every living thing keeps its colour; the human becomes, in sum, pure white. Every living thing lives in its territory; man wanders through the world. The origin of knowledge lies in the totipotency of our body and of the relationship to the world this totipotency allows, as cause and as effect: this omnivalence gives it a certain independence in relation to every niche, and this independence hastens this omnivalence. We live in the evolutionary sum of these places; we wander from one to the other; this is why we travel, decentred.

So, contrary to the vulgate on this point, we must celebrate the famous blows struck against human narcissism – the discovery of America, the Copernican Revolution and others – since the distance between the old lost centre and the new one makes our world larger. With every decentring, we leave a narrowness and increase our wealth in world. The knowledge living things have forever remains, as it were, a knowledge-of-something; our knowledge, firstly and lastly, has an encyclopaedia's round and ceaselessly swelling shape: the understanding of the world. The ancient celebrations of our body as a microcosm described this evident fact. Humiliated in the past by ten decentrings even though all of them were strengthening his fortune, the nothing-man, reduced to nothing, they say, radiates with all the devices that issued from his body; these devices end up in a totipotent circle whose circumference repeats the originary and fecund hole of his nullity. This omnipotency encounters the whole of inert and living things as an encyclopaedia-world of existents, actions and thoughts.

The sciences reread fables

Always dealt with in a linear way, either according to the time of a genesis or following the sensory channels taken one by one, the question of the origin becomes clearer in the global spectrum of this white light. In the body, in its gestures, in its movements, in its food …, there isn't anything in us that isn't always already of the order of totipotency; there isn't anything in which we don't show our omnivalence. For, at the start, a certain omniscience is exhibited in a dazzling way in totems and fetishes, part human and part animal, and in fables and myths, in which metamorphoses function to the utmost: we thus travel across every genus of living thing, as though each of us were imitating a species all by himself, as though, in sum, humanity were entering into every species. You, a bull, and me, a fox. In this sense, all our knowledge, itself universal, issues, like perennial springs, from fetishisms and totemisms, from fables and myths, always present among us as during the first days. Regarding the cognitive sciences, we learn things of the same sort in fables, in stadiums, in which trained bodies run and jump, with the handicapped, whose gestures and postures explore the possible even more in order to survive by compensating …, and in laboratories or via computers.

How, consequently, can we be surprised that in return our knowledge, in advancing, gives us the possibility of continually revisiting the entirety of this vital scale and, for two decades, through the function of reproduction? We have recited the authentically new and hominescent lists of Jean Rostand and Testart, intoned them, practised them since the dawn of time in other application sites: gestures, foods, hunting postures, dances, rites …, sports and cooking recipes, Amerindian, Aboriginal or Egyptian religions, Ovid, Aesop or La Fontaine …, the humble arsenal of knowledge issuing from the ancient or the exotic. Before the horizon of totipotency, nothing distinguishes the Australian Aborigine from the Nobel Prize winner in biochemistry. Sciences and humanities are rooted together in these strokes of hominescence.

Diverse diversities

Having passed into being a key word since Montaigne and Pascal, the notion of diversity today applies particularly to cultures or languages, as though it only affected the social sciences. The struggle against every kind of imperialism passed first through this demand; so we remember its lesson. But let's recognize that medicine, the veterinary art, agriculture, livestock farming and natural history have known and practised the diversity of bodies much longer than we have admitted the diversity of customs. The living is even taken to be the domain of the diverse par excellence. The livestock farmer doesn't know any cow but Antonia or Mean Girl; the doctor treats sick people not sicknesses; before anatomy arose, the body was spoken of less than the individual; we are speaking about it again these days thanks to subtle explorations and algorithms. So I am drawing a parallel between this vital mosaic that's fundamental and integrable, as I just said, and the cultural and linguistic mosaic of civilizations, non-integrable, perhaps. Not that the one can be taken to be the cause of the other – always the misdeed of linear logic – but that all of them together participate in a similar mosaic. So here, the diversity of languages encounters the despecialization of the organs of phonation and hearing the way the plurality of arts is in harmony with the hand's omnivalence.

A new humanism with the mosaic colours of civilizations and knowledge is drawn from totipotency. For the passage to the possible resolves the Liar's Paradox applicable to relativism and its multiplicities: if everything is relative, relativism is also relativized. Great Pan has returned.

Mosaic humanism

Since there cannot be any humans except through this omnivalence and this nonlinear mode of living first before thinking, a new humanism having the widest diversity comes about. The plurality of languages and multiculturalism cohabit with the diversity of living things, a diversity as sparkling in its variation of individual bodies as in its chain links of DNA, in which the diversity of languages is reproduced, as though in image, in and by an alphabet of simple elements.

Are languages, whether written or merely oral, so many varieties of a universal language the way each individual varies on the genetic code of every living thing? Here again, we didn't imagine that the diverse families of languages spoken in Eurasia could be grouped together, a union that's probable today; nothing, it seems, is standing in the way of a hypothesis that was still rejected as myth up until a recent date. Of course, the pixels change: carbon or nitrogen atoms, amino acid molecules, cells that are sometimes, them too, totipotent, the variable organs of evolving bodies, language consonants or vowels, religious rites, tools and customs, the divisions of space, the arts and the nine Muses, the disciplines of the sciences assembled as an encyclopaedia ... Yes, the little component stones change, but the form itself of the marquetry remains stable, from life to knowledge and back. Little science distances us from this mosaic and totipotent humanism, but a lot of science brings us back to it. Little humanism distances us from science, but a lot of it brings us back to it. Here, philosophical thought marvellously imitates the first of the fine arts, music, another pronunciation, harmonic this time, of the word 'mosaic'.

Equipped with their devices, the experimental sciences bring us closer to the integration, to the fusion of the boundaries separating the mosaic pieces (or to the connections between the foliations of the Riemann surface), whereas the social sciences, highlighting these divisions instead, like to make themselves the guardians of the marquetry. White sciences in the style of Pierrot; multi-coloured knowledge in the style of Harlequin's clothes. The fact remains that in order to metamorphose into an Australian Aborigine or a Gascon sailor, in order to appear as a WASP or a Kwakiutl, some animal must have this power. This power that is, once again, totipotent. This truly corporal power from which culture takes over the reins. The other living things change bodies and species through mutation and selection, according to their biotopes; this sometimes requires millions of years. We keep the stability of a certain somatic omnipotency and make customs, languages, rites and tools, symbols and representations of the world strongly bifurcate.

In a way, the powerful differences rightly noted by the social sciences and carefully preserved by them thus prove, *a contrario* as it were, human unity and its power of integration as well as the unicity of the world observed by the sum of bodies. Culturally differentiated, the body keeps this unity invariant, a unity that's not abstract but present in its flesh. Lazy, Pierrot sleeps behind the boisterous Harlequin. Under the paving stones, the beach; under the marquetry, whiteness.

Once again, finitude

Contrary to what Leibniz, Kant or the existentialists claimed, we discover ourselves to be without limits or bounds, in space, time, the real and the virtual.

Virtual, we are not real. Infinite, we are not perfect. The Greeks in effect but not without reason associated finitude and perfection: as an obstacle or a support, a bound defines a space, reacts to a force and permits thinking with precision. It gives meaning to History and a single direction to a project, assurance and clarity, foundation and exactitude, effectiveness of action and distinction of idea. We lack this drawing, this definition, this constant of force, this line and its goal. Virtual and infinite, our time continually bifurcates, as contingent, multiple and undefined as ourselves; it is integrated as well into the white of possibilities. If this omnivalence didn't exist, no one would understand why or how, unexpected, we can disobey, conforming so rarely to what anybody says about us, thinks about us or orders us. Except in cases of voluntary or passionate subjection, no one can predict our intelligent behaviour, nor our behaviour in love. Say who I am, and I immediately become other, an animal or hero of doing the opposite.

Neither finitude nor bounds increase anxiety but, on the contrary, dispel it because they define the threshold of a house, the wall of the garden, the borders of the tribe and of its language, the *non plus ultra* of dogmas and customs, the anvil of the hammer, the surface of the sphere and the foundation of thoughts. Finitude gives a sense of security. The open seas where the swell does nothing but announce the swell, the desert with recommencing dunes, the space of cosmonauts …, the vague, the open, the free ball of possibilities, in short, the absence of boundary frighten. The well-defined real erases anxiety; the virtual arouses and deepens it. We used to like living in networks, at crossroad points, separated by roads having calculable distances; today we are wandering in a constructible world without any markers or intervals. This recent change of habitat alarms us. The new house-world causes us to be afraid. Furthermore, everything depends on all of us; nothing could be more terrifying. For totipotency knows no bound, and this status worries us. Without definition or seamarks, we wander on the ocean of the undefined, translucent and floating phantoms.

Anthropology, theology

The only previous mention of this omnipotency I can find in History is in the theology of the attributes of God. Ever contrary to received ideas, the history of religion offers one of the most effective and profound motors of the history of science and even of knowledge in general: a tectonic plate whose slow advance gives rise to fearsome convulsions on the surface.

The French word *ordinateur*, which translates so well the English piece of nonsense 'computer' to designate a machine whose performance far surpasses simple calculation, also comes from medieval theology. In the middle of the Enlightenment, Diderot still mocked he who 'thought that the universal, ordering [*ordinatrice*] and first cause was good'. The Christian religion is acquainted with three '*ordinateurs*': God Himself as the universal and sole cause, Christ next as the initiator, incarnated into human time to put back into order what had fallen into disorder, lastly the priest who presides over rites and ceremonies, each of them proceeding from the preceding one, which confers on him his power and dignity. This so exact translation into French of the name of a machine that produces order contributes to making an attribute of God, the '*ordinateur*' [orderer] of the world, slip towards us.

This book sketches out the same gesture of translation for omnipotency. That totipotency designated by a religion and its theology as one of the divine attributes passes to humanity as its own possession, not in eternity but in temporal becoming and through a thousand virtually integrable contingent processes. Continually putting its destiny in order, a destiny to be chosen in real time, defines open human freedom, which today receives, in addition, the entirety of the means of choosing what its time has accumulated. Yes, we are becoming all-powerful; our mastery goes around widening to the global inert world, to space, to the length of the time to come, to the entirety of our species and even of living things, to the horizon of the virtual. This is why anxiety is choking our contemporaries, responsible in fact for future generations, and obligated, because of this, to construct their grandchildren's evolutionary house. This fear overwhelms them to the point of themselves wanting to live lives with zero risk, without realizing that the very thing that makes life worth the trouble of living is precisely worth the sacrifice of life and that existence with zero risk consequently has no value. And therefore, at vital risks, the human adventure worked for millions of years, chaotically no doubt, and the totipotency resulting from these contingent tasks suddenly appears retroactively to us with a dazzling clarity: it integrates the chaos of time.

But, long terrified of this result, we desired, with all our anxiety, to put totality back into the hands of God at the same time as omniscience. We hid this totipotency because it implies, no longer a theodicy, but an anthropodicy, I mean the appearance of the responsible party, now human, for all the evils in the world before a court. For we have deposited this power into our own bodies, our own intelligences and our own means at the same time as the atomic bomb, genetic engineering, the demographic explosion, in short, our integration in process of the world. By producing or receiving totipotency, our metamorphic destiny is therefore encountering omni-responsibility today. Caution: we are going to spend our time searching for guilty parties; this is the historical misfortune to come. We are even already indicting our own knowledge.

God himself formed or still forms the horizon of this destiny. We inherit His attributes. Therefore we find ourselves causes of or at least responsible for hominization itself, for the world we inhabit and make use of and which, consequently, we create, universally, in the manner of a continuous creation; totipotent, we find ourselves bearers of all the ways of doing this. Conversely, God inherits all the attributes that we formerly left to humanity: weakness, the life of the worried victim, wandering and persecution. Infinitely fragile, infinitely little a cause, less universal perhaps than even ourselves, we have every freedom to forget Him, to ignore Him, every faculty to abandon Him, to spit in His face, to stomp on Him, to condemn Him, even to kill Him without standing trial, and no one dispenses with Him without thinking that we will no longer ever be able to abandon or forget this totipotency that characterizes us and widens irrepressibly. Covered with opprobrium and spit, impoverished and abandoned, God now only inhabits the third and fourth worlds. In this strange crossing of legacies in which modern anthropology exchanges its concepts with traditional theology, there remains one and one alone that still escapes this exchange and before which we can no longer, we must no longer hesitate: for the entire lot of the inheritance has to be accepted or nothing at all. In becoming our own cause, the continuous creator of our world and of ourselves, in gaining the totality of possible powers to do this, it remains to become good. Providence, omnipotence and omniscience, these are the old attributes of God. So this is what, infinitely remote, we lack and remains for us to gain. The *Theodicy* is turned around: Who then must stand trial for having chosen evil rather than good? The Totipotent One, now condemned to forbearance.

Without mercy, the other legacies aren't worth anything. Yes, philosophy has gotten its fill of *sophia*, science and understanding, but, a strange half-measure, it hasn't yet begun to experience, with love, *philia*. Knowledge and totipotency cannot do without an infinite pity. Otherwise they become monstrous; otherwise the new God becomes fouler than every demon the imagination of artists had ever conceived or painted. A weak victim full of love, the overturned God I see before me seems more worthy of veneration than this new humanity if this latter robs Him of every attribute except the last one, goodness. A crowning glory without which, according to the cycle from just now, the preceding legacies would lack meaning, love manifests the ultimate sparkle of totipotency, for it 'pardons all things, believes all things, hopes all things, endures all things'.[10]

8 THE EVOLUTIONARY HOUSE

Existence, time and life

The thought or philosophy of being, ontology, as I said, yields its place to modal thought. We are not; we exist in the following square: subject to *necessity*, gravity, colds, hunger and thirst ..., we choose and incessantly undergo, among a hundred *possibilities* that are hoped-for or projected but sifted by as many *impossibilities*, a time, a future, a *contingent* existence. I could have not lived like this but can't reconsider my choices: from the present to the past, contingency turns into the necessary, which nothing will change anymore; I don't know my future, unforeseen in sum, onto which I press with hope and preparations, but don't delude myself with unachievable dreams; sometimes, however, by chance, a miracle ... From the present to the future, the necessary turns into the possible; filtered by the impossible, it arises as contingent. These four modes sculpt existence in real time, and time itself to boot. Thought, for example, by chaos theory or percolation theory, time enters with them into this square of modality.

They also mould the living. Compatible with the inert and made up of its elements, subject therefore, like it, to gravitation and thermodynamics, living things are born from the laws of physics and chemistry, from the realm of necessity. The life being born opens up a thousand possible paths: Who could have guessed, before the event, the Cambrian explosion, during which so many species emerged? But, among these multiple choices, constraints forbid, with their impossibilities, some path that stops or bifurcates; they even kill some genus, which goes extinct. Species percolate across the mutation–selection or possible–impossible barrier. Lastly, every living singularity could have not been born and lives from contingency. Life shoots up from the necessary to the contingent by crossing the hedge built by means of possibilities filtered by impossibilities. The square of modality thus summarizes the biological, classical and modern sciences and gives them their letters of philosophy.

Life haunts, in sum, the same square as existence.

Biotechnologies

From theory follows practice. For biotechnologies work in the same square from which humanity and every living thing as well as their existence and their time rise and gush forth: already following and mastering the necessary laws of physics and chemistry, in sum of the inert, biotechnologies attain, via the genome, a spectrum of possibles, filtered by the impossible, that is to say, ending in death, and thus create contingency.

Our existence used to be subject to these modalities in the same way as every living thing; they named our state and designated our destiny, fatal, fragile, tragic and illuminated with hopes, suffering from constraints and ills; we partly master them today. Why? The Baconian adage that nature must be obeyed to be commanded only concerned the necessity of the inert and its linear logic; this simple action–reaction schema imitates the tactics of those combat sports in which certain gestures profit from the adversary's attacks in order to turn them against him by utilizing their own force: the reaction suddenly inclines and reverses the action. Having entered into a logic in which bifurcations bush out around this line, we enjoy a wider space in which we can turn around, in which we see paths outline themselves that are numerous enough for as many choices to open up.

Our laboratories are built inside this square, as though it were a question of their walls. At one of its vertexes, the impossible is therefore found, death, in other words selection, natural since Darwin, agricultural and therefore artificial since the Neolithic, artificial and non-agricultural since a few decades; the mutations of the genotype and its relations to the phenotype occupy the vertex of the possible; the inert, the laws of physics and biochemistry lie at the vertex of necessity; at the vertex of contingency triumphs, fragile, a singular living being. The various Darwinisms installed the possible–impossible diagonal under the double name mutations–selection; like a kind of filter, this diagonal traverses the passage of the other diagonal that goes from necessary biochemistry to the emergence of a contingent singular living being. The one crosses the other the way the élan vital shoots out across a hundred constraints. Along one of the sides of the square, material necessity pilots the combinatorics of possibilities and codes its algorithms, while forbidding, from the other side, exceptions to its laws and sanctioning them with death; in other words, the 'left and lower' half of the square remains of the realm of biochemistry, advancing towards the originality of living beings; the other half, from the other side, 'upper and right', of the diagonal, opens up the realm of biology and the natural sciences, of evolution and contingency, of the singular unicity of the living individual: from the possible to contingency new things arise that are planed down, from the side of the impossible, by eliminations ensuring their excellence. Thus, in binding together what ancient metaphysics called matter and life, the square of modality summarizes the genesis of life. While this schema unites and links the execution of material laws to evolutionary functioning and the naturalistic acquisitions of the nineteenth century to the biochemical discoveries of the twentieth century, while it simply explains and allows us to understand, it also opens up a set of practices.

Metaphysics

One of the stumbling blocks of traditional philosophy consisted in the fact that an unshaky line opposed mathematics to existence, necessity to contingency, and Aristotle, who gave full reality to individual substance, to Plato, who granted it to the Ideas. General laws don't attain singularity, which remains above or below. Or if they do reach it, we must say how and show the path step by step. Leibniz first drew and formalized this line by opposing, precisely, *mathesis universalis* to the monad, a singular substance, but immediately diagonalized it by crossing it with the other line, the one that goes from the possible to the impossible. He deposited the entirety of the schema into God, at the crossing of His understanding and His will. The lower and left half of the square represents the former, and the other half, upper and right, represents the latter. The understanding contains eternal truths and possibilities, while excluding 'incompossibilities'; the will chooses the best from among the 'compossibilities' and creates it, brings it into existence or makes it live in its factual contingency. At the same time, in the technological realm, the same Leibniz didn't neglect to bequeath to us combinatorics, binary numeration and algorithmic thought, whose course happens precisely step by step, without taking into account the repetition and perfecting of a calculating machine in Pascal's style, all of them operations and mechanisms that lead this abstract square to practice.

Yet biotechnologies today imitate this very process that Leibniz described as though he had been witness to the *fiat* of creation or that God had installed in His glorious design. Bossuet reproaches him, in fact, for his 'mad anticipations of triumphant life'. Our recent feats improve on and apply the operators he prepared – algorithmic code, machines – in order to realize what seemed proudly inaccessible to Bossuet. Some intermediaries were necessary of course: Buffon, Lamarck, and then Darwin, Mendel and de Vries first translated the possible–impossible diagonal into evolutionary terms better adapted to living things: mutations–selection; Schrödinger and then Turing expressed the necessary–possible side in information terms and then in algorithms better adapted to machines; and again Darwin and then Bergson described the emergence of contingency among the spectrum of possibilities.

These translations prepared, in theory, practices in which the recent practices of chemistry and the older practices, known to Lamarck and Darwin, of agriculture and livestock farming meet in and by the same square. Seeing only phenotypes, the farmer killed ill-come animals and burned ill-come flora in order to preserve the best varieties for reproducing or feeding us; the mutations whose posterity wins out in number and quality, says Darwin and his successors, survive, while the other ones die. Reaching the genotype, his biochemist successors discover sequences of elements there: so mutation becomes a permutation or at least a rearrangement. The natural process of the origin of species and the artificial act of the livestock farmer turn into DNA sequences and other long molecules, therefore, at the same time, into a combinatorics, in long rational chains, entirely simple and easy, which the algorithm specialists are in the habit of using to attain the prodigious complexities characterizing living things.

Combinatorics: Everything is number

Starting from this combinatorics, whose art Leibniz discovered at the age of eighteen, starting from this algorithm of some notes or other, the linear logic of traditional science was opened up. A sequence of elements, notes or numbers, in which substitution plays, replaces Descartes's deductive chain. Since the substitution of one note for another or of one number for another corresponds to a completely different chain, the single necessary line becomes erased in favour of a multiple bouquet, unfurled by these substitutions. So, in science itself, the possible arises, but also in the technologies and the arts: this is how it is with a melody that a change of a single note would transform; this is how it is with fugues and counterpoint. Did you know that on the day of his death Johann Sebastian Bach left a copy of Leibniz's *De Arte Combinatoria* on his table? The endless variations of *The Well-Tempered Clavier*, *The Art of Fugue* or the 'Chaconne' for violin were inspired by these permutations, so close to the mutations producing, in nature and by our technologies, living variations.

Already Leibniz, the combiner, claimed that 'while God calculates, the world comes to be'; already Buffon attempted to reckon the years of the Universe; already Mendel counted pea offspring by varieties; already Darwin assessed the number of those who were descended from the fittest; already Turing decoded by enumerating …, all the way to the combinatorial explosion, whose enormous numbers account for living things. Thus Perette, in her head, was already counting on her calves, cows and broods.[1] Yes, practice came from the fact that everything here reduces to the numerical. Everything is number: from the necessity of the measured, weighed, calculable laws to the combinatorial spectrum of possibilities and their filtering by constraints of all orders, all the way to the original and contingent unicity of the living thing, individually nameable by an exclusive code; everything is number. And these numbers make numerical practices possible.

A unitary philosophy

This first opening up the combinatorial sequences bring about on the single line of the traditional logics lastly gives the sciences a broader field, one which prepares this square of modes, where the necessary welcomes the possible, the impossible and the contingent. A rigorous knowledge and exact technologies of contingency and possibility no one suspected would one day enter into the realm of calculation or of mastery, of *mathesis* or of *praxis* arise there. Of course, in Leibniz's time, living things were still far from combinatorial algebra, with the exception that livestock breeders and farmers had been crossing varieties of fauna and flora for a long time, imitating God, combiners without knowing it. Voltaire's famous advice to 'cultivate your garden' is found, of course, in Leibniz, whose permutations and optimism in the choice of the contingent varieties created by the bouquet of crossings was no more understood by the naive ironist of the Enlightenment than a contemporary would view biotechnologies as the faithful keeping of the promises of this gardening or as the human advent of a transcendent theology. Yes, today biotechnologies are completing the expert labours of the ancient farmers with practices stemming from a metaphysics Candide mocked.

This metaphysics however has founded our house. For today, not only existence and time, the fragility of the individual, whose drama is lamented or extolled by works of art and the humanities, but also the sciences and technologies, our omnipotencies, as positive as they are tragic, but lastly the very unfolding of history, understood as sweeping a bouquet of possibles, desirable or dreaded, towards the unforeseeable contingent through the various filters of the necessary and of the impossible – in short, philosophy as well as the encyclopaedia that serves as its support – inhabit this square of modality.

Do we need another house-world?

*

The anthropology of cloning: Filiation

Who will inhabit it? A new and old family. Biotechnologies attain cloning, of cells, plants or animals, with the cloning of humans posing a number of problems. Yet, here and there, twins, even quadruplets appear, each of which, to my knowledge, has its identity without any mystery; culture distinguishes organisms that are naturally similar without any difficulty. Among the arguments raised against this artificial reproduction of an identical copy, the best of them insist on the major role played by filiation in the genesis of symbolic systems. Anthropologists affirm, and they are right, that without the decisive mark of a place in a family or its equivalent, reprogramming the human beyond animal life fails. In order words, through symbolic systems, filiation inaugurates culture. But cloning disrupts these bonds: we no longer know who is the son or the daughter of whom. How are we to resolve this question?

Specialists no longer know anything but Oedipus or the Nambikwaras. Who wouldn't be pleased that the distant in time and space should become near and familiar? On the other hand, the complete forgetfulness of the events that opened up the Christian Era, our era, weighs so heavily today on contemporary work, in which the neighbour distances himself as much as the distant person comes near, that I see myself compelled to remind the reader that this era began precisely with a treatment of this question, a treatment so original that it proposes solutions right where we only come up against problems.

The holy family

So let's consider the Holy Family: Jesus wasn't born of Joseph, called the adoptive father; the son of God the Father, of course, but it is written that his mother conceived from the Holy Spirit; Scripture calls him, as well, son of Man. 'Which of the two do you want me to release to you,' Pilate cries out during the Passion, 'Jesus or Barabbas?' The three Synoptics here are in agreement with John (Matthew 27:16; Mark 15:7; Luke 23:18; John 18:40) in relating the decision to crucify Christ and to free Barabbas. Yet this latter name signifies: the son of the father; therefore the one who was distinguished from him and died wasn't the son of the father.

Nothing could be more common than a child born of an unknown father, run away or dead during the pregnancy; by convention or by recognition, paternity doesn't know any natural law. Quite the contrary, maternity is related to the universal laws of nature, which tolerate no exception: we don't know of any child without a mother. But Mary's virginity introduces a rupture into this law and, as it were, a rarity. But how and why was that oxymoron 'a still virgin mother' invented? In order for maternity to be absent, partly, the way filiation and paternity were just now, completely. The father is not at all the father, and the mother is less the mother, a more extraordinary thing. Better yet, calling Mary Mother of God, *theotokos*, as a certain council did, amounts to giving this woman the title and function of mother of her father, an even more extraordinary thing. The adjective 'holy' in the expression 'the Holy Family' therefore signifies that this family undoes carnal bonds, biological bonds, social bonds and, as has been said, structural bonds: each in its way, the father is not the father, nor the son truly the son, nor the mother absolutely the mother; no more blood relations.

Even before the Age of Enlightenment, these rarities made rationalists laugh and, more recently, the social sciences enjoyed comparing this history said to be false to the strangenesses related by myths. Deeper and gentler in their practices and beliefs than the learned in their reasonings, the people aren't mistaken here and have, from the origin, recognized as Christian every woman or every man who calls someone brother, father, mother or sister who, precisely, is neither his sister, his mother, his father nor his brother by blood but who has freely wanted to become such in an entirely different register. This recognition defines Christianity as the deconstructor of the bonds of blood relation.

The laws of the sciences called hard describe physical necessities, while human laws, they say, derive only from conventions. Except for one at least and no doubt: no one can fail to be born at the opening of a vulva, the natural site from which every culture flows. The Greeks even called it the place, the reference point, the 'there' par excellence, the 'being-there', in sum. On, for or against this biological law, without any known exception, are constructed the structures of the family and the cultural and juridical laws of kinship, said to be by blood. So Christianity substitutes the paradoxical freedom of choice and the possibilities of love for this physical necessity from which no one escapes. The adoptive dilection that adds them up composes, at will, structures of kinship, including maternal kinship. The two dogmas of the virginal conception of the Word and of the Immaculate Conception discover the origin of culture in nature, and therefore of freedom in the face of necessity.

The inheritance of the covenant: The juridical point of view

All this bears witness to, at least, without reducing to it, a matter of inheritance. It has to do with the first Christian law. For without adoption, there can be no inheritors but those who are born to it, to the exclusion of the others. Christianity opens up the inheritance of the Covenant, reserved by the text subsequently called the Old Testament for the chosen people alone, to the universality of the human race, *omnes gentes*. The New Testament brought by the archangel Gabriel opens this legacy up to everyone. It therefore substitutes inclusion for this exclusion or, better, generalizes a singularity. The Scripture of the modern era writes the testament for all people of all nations, universally. How can this happen? Against genealogy and by adoption.

From its origin on, Christianity has undone engendering and reprogrammed it. All humans can, if they want to, become the adoptive sons of God and enjoy the Covenant; the previous examples given to them by canonic theology all bear witness to the deconstruction, by adoptive dilection, of biological, carnal and blood family relations. 'Woman, what do you and I have in common?', Jesus says to his mother during the Wedding at Cana (John 2:4). Joseph, Jesus and Mary, father, son and mother without truly being so by blood, make up the family of adoption, a word constructed on freedom of choice and which juridically translates, but in plain words, free and holy universal love.

For Christianity, adoptive dilection plays the role of being the single – abstract and formal – elementary structure of kinship. Free and chosen love becomes the only atom of relation, including in the most elementary connections of family. Hence the bursting of familial relations and their new universal and rational character in the guise of strangeness. In addition, the ordinary elementary structures normally define local cultures; their deconstruction, the possibility of choice, the introduction of freedom into the necessity of blood open humanity, for the first time, up to a universal. Said strangeness becomes the rational condition for this universal.

Bloody death through violence and blood-related family formerly opened the two great factories of ancient myths: the Passion and the Resurrection of Jesus Christ closed the first one, and the Holy Family destroyed, from top to bottom, the second one. It installs null, free or formal structures of kinship: thus Christianity deconstructs the very condition for myths. Far from repeating them, it destroys them. Nietzsche wondered why we no longer invented them: because the deconstructive critique implied by Christianity is functioning, or was still functioning. They reappear – we have seen examples of this – with polytheism as soon as the New Testament becomes erased.

Immediate application: its familial schema resolves the problems posed to the social sciences by filiation on the occasion of cloning and biotechnologies. For 2,000 years we haven't cared about knowing who the father, the mother, the son or the daughter of whom is since we can decide this at will, depending on whether we love this man or that woman in God. Better: this is modernity; it frees us from myths, whose ferocity flows twice from blood. Are the sciences and technologies aware of being the faithful executors of the New Testament?

A return to what Gabriel said: Neither nature nor culture

The single adoptive structure of kinship frees kinship from the physical and biological fatality of bodies and blood, in other words, from natural necessity. But it doesn't become, for all that, cultural since the free choice of kinship bond doesn't depend on the languages or the conventional laws of a given society; for no one is stopping you from choosing a father, sister or brother in other, distant, collectives. Who doesn't see that all racism is eradicated by this radical deconstruction of the kinship structures founded solely on the bonds of blood?

The Lord's messenger announces to Mary that she will conceive from the Holy Spirit. This word designates this engendering, neither natural nor cultural, as spiritual. What does this latter term mean? The sum of the two other words or of their negation: the addition of the supernatural, and, even though no one uses this word, of the supercultural, understand by this the universal. This is how Christianity reconstructs the symbolic system. The universality of the Spirit has its source in this sum.

The announcement made to Mary signifies that starting from the interval between 25 March and 25 December of that year, time would no longer be reckoned according to the genealogy or the tree or the structures of kinship: outside blood, kinship will be called universal. One of the Gospels begins with Christ's genealogy (Matthew 1:1-16), the sequential description of the Tree of Jesse, while, on the cross, the dying Jesus hands down, not life via blood, but precisely this choice. Before his Passion, the people had already distinguished him from Barabbas, Son of the Father; at the very moment of his end, his final words were addressed to Mary, his 'mother', 'Woman, there is your son', and to John, his preferred disciple, 'Here is your mother' (John 19:25). At the moment of his expiration, the one everyone distinguished from the Son of the Father handed down to his beloved disciple, by literally stating it, the New Testament: the mobile, formal, free and delightful adoptive bond.

9 THE SECOND LOOP OF HOMINESCENCE

The history of heat: World-objects

Here is the world opened up by our global body; here are its walls and inhabitants. How do we construct it in practice? This is how.

When we heat liquids and gas, vaporous mixtures expand starting from centres in every direction randomly. Recent ice cores from glacier ice sheets can in fact date, to the year, the beginning of the Bronze Age thanks to traces of the first discharges released into the atmosphere by the archaic furnaces of the Middle East and scattered everywhere in such a way that they were dragged by snowfall into those high latitudes. Who would have thought that globalization began from prehistory? Generalized, spread by the Industrial Revolution, heat technologies accelerated the rise from the local to the global. Our know-how has devoted itself, from a fairly recent time, to fashioning world-objects. A satellite, for speed, an atomic bomb, for energy, the internet, for space, nuclear waste, for time … These are four examples of world-objects. Can they still be called objects?

What is an object?

So what is an object? In the literal sense: 'that which is thrown before'. Do world-objects lie before us? The global dimension characterizing them does away with the distance between them and us, a gap that in the past defined, precisely, our objects. We inhabit them like we do the world. Do we call our houses objects?

Traditional man-made things, tools and machines, form sets having local radii of action in space and time – the awl pierces the piece of leather, the sledge hits and drives in the stake, the plough cuts the furrow … – and define, in all, an environment on which the experts who use them work. Such a dividing up of the world into quasi-corporate, even specific, localities makes a philosophy of mastery and possession possible since we then know how to define what we dominate and how we do this, just like the animal species do. This stable division contributes to defining and clarifying the medieval notion of object, *ob-jectus*, that which lies, at a middle distance, before the body and its strength, in our interventions and thoughts. Held by a subject, a technological object acts on other objects, sometimes on other subjects; all these elements remain within a narrow spatio-temporal subset, one relatively invariant in time.

Begun with heat technologies and the quantitative increase in these world-objects, globalism little by little forms a new universe: technological, physical, as we can see, human and juridical, as we shall see. Can the things that constitute it still be called objects and the people who use them still be called subjects? Are our communication networks objects? They have neither the presence of objects nor probably the reality of objects since channels and fibre optics transport numbers, symbols and virtualities. We inhabit them instead.

Dependence and possession

Lastly, as much as it is possible to master given places in short times and to make ourselves possessors of them since in the end property can only be understood with the occupation of a quasi-biological delimited niche, we are ignorant of the ins and outs of a global mastery of the world. Yet, devoted to difference, current philosophies all say nothing about the categories of totality, so difficult to handle, since we only know how to define with rigor and exactitude in the local, that is to say, for objects, in the medieval sense of this word. The Cartesian adage of the possession of nature doesn't define the conditions for the mastery of such a vast 'object'.

This same recommendation for mastery is inscribed, on the other hand, in the slow historical displacement of the old Stoic division of things that depend on us and things that don't depend on us. And, once again, what 'things' and what 'us'? In this second act, a Cartesian one, those 'things' that formerly did not depend on us suddenly depend on us, and more and more so; we sometimes even live the certainty, not always illusory, that everything depends on us; but, third act, we are beginning ourselves to depend on things that depend on acts we undertake, things given rise to, unleashed, in any case born from our actions, like a new nature. After the first Stoic division, after the second Cartesian mastery, a spiral follows in which mastery and dependence inter- and retroact, in which, by mixing, the outmoded subjects, formerly solitary, disappear along with the old outmoded objects, obsolete because localized. Is it the change in status of objects or of things that commands the change in the regime of dependence or possession? Already thirty years ago, I had written that after the mastery of the world, today the master of mastery must follow.

World-objects put us in the presence of a world that we can no longer treat as an object: objective, assuredly, and in this way we avoid all animism, but not passive since it acts, in return, on the global constraints of our survival. We must think such a return.

The second loop of hominescence concerns the world

Less and less objects, more and more world, world-objects lead us to a world which isn't an object like the objects of the world. We don't know this world in the same way. We don't know how things are with it; we are barely beginning to understand it, and this knowledge differs from the knowledge we have of a delimited object. We are barely beginning to undertake actions on it, and this practice differs from the practice by which we acted on those old objects. We are barely beginning to know who acts on it or knows it. Who should be called 'us'?

The first loop of hominescence rebounds on our body: the environment we produced and onto which we had externalized its evolution influences it in return so as to make it bifurcate towards another evolution. The second loop concerns the world that we shape starting from these world-objects. Seventeenth-century philosophy said we were *naturata* or natured; we are becoming *naturans* or naturing, as I just said: we cause to be born, in the etymological sense of the word, an entirely new nature, partly produced by us and reacting on us. We became the humans we are from having technologically sculpted our environment, our own house, in order to protect ourselves; now made up of these world-objects, this evolutionary house even acts on the world as such to produce this new nature, an astonishing mixture of the world as such and of this up till now protective environment.

What Hindu, medieval or ancient thinker, what propagator of the Enlightenment, what recent rationalist would have guessed that one day, without sorcerers or magicians, we would be unleashing thunderstorms and warming the planet? Faced with our world-objects and through them, the world is becoming technologized and culturalized.

Concepts and images

So philosophy is re-examining its old concepts: the subject, objects, knowledge, action …, all of them constructed for millennia under condition of localities whose divisions defined, additionally, a subject–object distance along which knowledge and action played. The measure of this distance conditioned them. Divisions, proximities, distance, measure …, these finitudes conditioning our theories and practices are coming undone today, during which we are moving on to a larger theatre and losing our finitude. Old and timid categories of totality, like being-in-the-world for example, are entering at the same time into objective knowledge and technological action. So they are moving from metaphysics to physics, from speculation to action, from ontology to modal logic, from contemplation to responsibility.

We have acted enough on things; we have attempted to examine its objects; it is time to know the world or to discover nature, not in the ordinary senses, but, I repeat, in the pure etymological sense since nature is being born before our eyes, in and of our hands, completely new for us, our globalized knowledge and our globalized acts. Result produced, nature itself returns as a condition for knowledge, action and even survival behind the new subjects, who are immersed in it, as soon as these new subjects act on it.

Reversal of the subject–object relation

Thanks to photographs taken by astronauts, we see the entire Earth, newly born. This vision has nothing to do with the old perceptions, which supposed, behind their forms and as background, that never-seen entire Earth. Being-in-the-world had never eyed the world. Through the internet and email, we communicate immediately with the same globality, constructed and given. Through our technologies and their discharges, we act on this globality, its climate and its warming. As soon as we act on it, it changes and causes us to change; we no longer live in the same way, wagering on the consequences of this reaction on our survival.

On the other hand, in the global system of knowledge or of acts, for the world or for humanity (a new 'we' in formation in the same way), sometimes a profit in one place corresponds to a deficit in another: global warming can be translated, here and there, as a glaciation. Local actions and thoughts could not see it. The global often resolves itself across distributions, difficult to predict, between losses and profits. Thus new communities appear, and a global public opinion forms, which are likewise homogeneous and nevertheless mosaic. Therefore humanity corresponds to the entire Earth, a humanity no longer abstract or sentimental (at least and in the past), or potential (at most and recently) but, grey and multi-coloured, actual and soon having real effect. Of course, everything depends on us, but we are barely beginning to understand what both this everything and this we signify. Through the sciences of the world and of the human, lastly we are reconstructing a new grand narrative, which, concerning all humans and the world in its entirety, gives hope for a decentred humanism, for the first time authentically universal. This we and this everything interact by reversing their roles.

Victims of our victories, we are in fact becoming the passive objects of our actions as subjects. Correlatively, the global object reacts to our acts, like a partner subject. At international meetings, what the officials sent from their states say about the warming of the Earth, for example, is less important than the awareness of it. This awareness of course is due to the new technologies, but also to the appearance of our common vessel, which has suddenly become, through its unexpected reactions, the partner of our acts and our worries. What then are our representatives increasingly talking about? About this global Earth. In the past brought into subjection by us, former subjects, itself become a subject of law in *The Natural Contract*, the Earth today is a political subject, a predictable thing. Stemming from the hard sciences, the Earth enters into History. This is also what I mean by hominescence.

Pollution: The cost of things

Classical philosophy never calculated the cost of knowledge, thought or actions: it prejudged them to be free. It lived in a world light with grace and the given. But as soon as work appeared, everything passed through the martial law of cost. Of Greek and Roman birth, philosophy presupposes slaves who pay with the sweat of their bodies for the freedom of thought and action of those who live in leisure. This projected cost has been calculated beginning from the Garden of Eden's exit gate, where the question of evil, labour and pain itemized its calculations. The productivity of work never rises to oneness because waste, refuse, mud and grime always exist, those objects excluded from metaphysics by Plato in the *Parmenides*. As long as there are slaves and serfs, as long as work remains cold and local, costs pass through profits and losses. As soon as heat enters into work, the productivity of the fire-based machines must be calculated. And as soon as world-objects function, the cost becomes commensurable with a dimension of the world: and, for example, the sea rises.

On this point, we could have taught the politicians that a collective doesn't behave like a person or even like a family. Just as a city or a nation represent wholes that aren't reducible to narrow groups, so the world is an object that doesn't have the same status as a local object. Changing scales implies a change of laws. In an even more archaic precondition, certain religious traditions announced, in the global, instances of such feedback. The early scene of the Flood, for example, which is present in many traditions, may describe a certain physical marine transgression, but it talks too much about peace, doves and olive branches not to be consciously warning that our human rivalries can put the planet and life, in their totality, in danger, a totality marked by the universal rise of the waters, a risk warned of by the gathering of the remaining animals in the Ark. Far from talking about guilt or moral interdictions, these scenes seem to warn that a certain global end turns us into objects that are dependent on our free acts as subjects: the sea rises.

Subjects, objects, knowledge

Let's return to the things themselves: for the linguist as for the historian, cases [*causes*] precede things, and the first known subject designates the subject of law; a contract precedes knowledge and act. The French word *chose* [thing], employed to cut out an objectivity, derives in fact from the Latin *causa*, a juridical term suitable for designating what is at issue in a trial or the trial itself. At the origin, *chose* designated that which is debated, the decision of a tribunal, that for which there is a contract. Knowledge of a *chose* ensues from its establishment by a legal authority naming both an agreement and its object. Likewise, English's 'thing' derives from a Germanic legal term. Therefore, in our languages a social contract always accompanies the emergence of a thing: whether the thing constitutes the group or the group the thing, we will probably never know which preceded the other. In any case, an objectivity appears at the same time as a collective, and this appearance takes place under legal conditions.

Likewise, the first known subject designates the subject of law. Consequently, *The Natural Contract* deals almost exclusively with this question: Who has the right to become a subject of law? The history of law shows the progressive universalization of the right to become a subject of law: slaves long ago became so, children thereafter and women much more recently, a decision whose late date brings shame on humanity. We shall reason tomorrow, in relation to nature, with the same shame as today for having taken so long to understand that our companions should have become our partners at least since the foundation of the world. The entire question deals with the status of subjects first and then objects. It seemed crazy to some to propose a Contract that would bind an object and by which an object would be bound: might as well make a horse a senator or nature a harsh mother.[1] Poetry or madness. To my knowledge, the same criticisms were directed at Rousseau since his *Social Contract* was never signed in known or knowable History by any man or collective, and since it designates, for the philosopher, the *conditio sine qua non* or transcendental condition for the formation of societies. Bacon could have been criticized in the same way: Who is commanded, who is obeyed in his famous adage that nature, to be commanded, must be obeyed?

Objects become transformed by globalization in the process in which action and knowledge increase towards the universal; the objective status of the collective varies since, formerly active, it becomes in return the passive global object of the constraints of its own actions; the status of the world-object varies since, formerly passive, here it is, in its turn, active in return, and since, formerly given, it becomes our de facto partner. I shall define how more precisely. But before doing so, we can no longer describe the scene of knowledge by means of the medieval couple subject–object: the words themselves change, along with their relation.

With regard to this relation, I don't know of any knowledge that doesn't also begin with legal conditions, whose impact in the history of science grows at least as fast as globalization conditions. For all knowledge requires an agreement or consensus that only de facto authorities and authorities by law take it upon themselves to establish. Teaching and research make one appear before examining juries, competition juries, prize juries or publication juries. Before saying that anything is true, false or probable, even before saying that this or that is or is not an object of science or non-science, some adjudicating authority deliberates and decides this during a trial in which the opposing parties are largely heard. Subjects of law say the law of objects.[2]

Yet today we have to think a new object, one surpassing by far the status of local objects, since in certain respects we are becoming objects of that which we don't even know if it is an object; if we treat the world like an object, we condemn ourselves to becoming, in our turn, objects of this object. In order to think this new situation, we therefore need to return to the original juridical gesture: this new object emerges to thought by a new Contract, one that establishes both this new global object and the new global group that thinks it, that acts on it, whose debates make it appear, whose actions make it react and whose reactions condition in return the very survival of the collective that thinks it and acts on it. For more than twenty years, we have spoken about nothing but it, debated about nothing but it, done nothing but set up the foundation of what I have called *The Natural Contract*.

Knowledge and exchange: The given

I have promised to speak about the partner. The question of the relationship between knowledge and its object has never been thought within the framework of exchange, as though it remained understood that the active subject took a piece of information the passive object gave to it. For the use, in philosophy, of the word 'given' reveals that the objective or exterior world gives and doesn't ask for or receive anything in return. As a result, the bond of knowledge becomes that of a parasite, such as I studied it in the book bearing that title. The subject takes everything and gives nothing, whereas the object gives everything and receives nothing. Free, knowledge can then be coupled with actions that would be no less free of charge. The active or technological relationship to the world exploits it, that is all. Parasitism or predation: we didn't know how we were behaving. What seems normal, usual, ordinary in knowledge turns into scandal and abuse in exchange. But if we start to know through juridical processes, a certain justice must take place in the exchange: hence the necessity for a Contract.

All pedagogy consists in making the human child into a symbiont or partner of a balanced or equitable exchange starting from the parasite he couldn't help but be at the origin. He will take, of course, but he must give in return. In a way, he ends up having to sign a contract of exchange with those around him as though he started out in human and civil life by learning an unwritten law. He very often falls back – alas! – into the quasi-animal equilibrium of the parasitical relation, taking as proof of this the fact that a donor rarely gives to a donee but almost always to a parasite who rises to intercept his gift. All pedagogy begins therefore with this Contract. Consequently, we must educate scientists, technicians, politicians and users the way we have educated our children since the origin of all pedagogy. We are becoming, belatedly, adults of knowledge and action. The knowledge relation is changing today through the demand for symbiosis with the new object.

Before knowledge, exchange; to make exchange equitable, a Contract is necessary. Knowledge begins with the law whose laws precede every discovery of laws; likewise technological action begins with the law of exchange. Then the symbiosis of the global world-object and the global human race-subject begins. Through this loop of hominescence, *Homo universalis* is born.

10 WHO, *EGO*?

Land and sea, being and relation

In the rural France of yesteryear, communities, on average 30 km from each other, succeeded each other according to when the carriage horses pulling the stagecoach had to rest or feed. The old departmental divisions in France retain traces of this web, woven with administrative centres and with sub-prefectures, sub-woven also with the principal towns of cantons.[1] Doubtless one could go by foot from farm to farm according to a denser lattice of local commune roads and dirt paths. With regard to the big capital cities, they required journeys from prepared post house to prepared post house. Likewise, the scattering of ports along the coasts of the Mediterranean or of Atlantic Europe depended first on the sail and the oar, next on the propeller and steam. Distant, the inhabitants of Copenhagen and those of Bordeaux became closer by sea; quite the opposite, neighbours, from La Rochelle to Sète, didn't know each other owing to the ground: water brought together those separated by land.

Running from inns to caravansaries and from landing stages to quays, the means of communication therefore contributed to drawing the maps of the habitat. How long did the ascendancy of the horse and the sail over the distribution of peoples last? Do we have any traces of a distribution prior to the domestication of pack animals, before agriculture and navigation, before the epic of Gilgamesh, before the *Odyssey*, which teaches so well how to distinguish men according to whether they swing an oar on their shoulders or a grain shovel? At what respective distances did tribes live when they were wandering just where the hunt took them and following the ripening of the fruits to be gathered? The nodes or crossroads of these diverse networks continuously vary according to the speed of transit on the roads. Existence depends on relation.

Hard and soft, slow and fast

Railroads and steamships restructured these maps, whether land maps or sea charts, but for a short time since, suddenly, the aforementioned means of communication multiplied. Technologies went from a single innovation, the horse or propeller, to several at the same time: from mononeism to polyneism. Cars and their freeways, air routes for propeller planes and then jet planes, the telegraph, landline phones and mobile phones, over-the-air and digital radio and television, the fax, the internet … are all suddenly jostling one another and in their turn weave several new webs, superposed or sometimes united. So the sites of convergence fluctuate, the metropolises swell, the medium-sized cities depopulate … Here, too, for how long?

For these new weavings must again distinguish the hard and the soft. We transport sugar and sand by barges or heavy trucks, and messages by optical fibre. Of course, we inhabited according to the flows of food or construction materials, hard, to erect our walls and provide food for our tables but also, without always realizing it, following the transmission of news, said to be soft. The former group prevailed over the latter, for while no one can survive without bread, everyone can go without knowing what his neighbour is doing. Our contemporary youth have trouble understanding a space where information circulated as poorly and as slowly as a hay cart or a rubble stone dray, always at risk of getting stuck in the mud. Up to and inclusive of the Second World War, we lived in a relatively heterogeneous and anisotropic world in which messages rarely arrived at their destinations, and most often distorted. We were ignorant of everything that was happening beyond the horizon, whereas the contemporary house is camped in an expanse that, from an informational point of view, is in principle and supposedly homogeneous and isotropic. Consequently, historians today have a tendency to believe that, yesteryear and yesterday, anyone could have had, as today, a global, almost historical, view of events. No, we lived and ate in isolates up until this polyneism of instantaneous means of information occurred. Sharing History in common was born at that time, not before when History remained the privilege of the great. Just as we used to have a tendency to forget the networks of soft messages, so today we take hard roads and routes less into account. Is humanity now only going to eat signs and take shelter in its sites?

In addition, we must distinguish, under the heavy sand, the slow barge from the light letter that flies, lightning fast. When signals instantly propagate, communications become detached from their space and time conditions: habitation places will no longer depend on them. Tomorrow our houses will be distributed randomly or as we please in space, which will have already lost, on top of that, several hard and heavy traces of agriculture. Reaching their maximum, their optimum, relations become established without condition and seem to disappear. We then understand that what mattered above all was their difficulty, their rarity. So let's correct our ideas: existence depended less on relation than on its uneasiness since it had to pass through obstacles. When there are no longer any mountains, there are no longer any cols; when there are no longer any cols, why build mountain refuges? Without any fords, why have inns? When these constraints disappear, why should one build one's house, live, sleep and eat according to the thresholds crossed by roads? My native town sleeps at the elbow of a river diverted by a hill: this isn't easily passed. These constraints having vanished, we will no longer live, not only in the local but in these singularities of space and time.

The neighbour

Light and instantaneous, the new communications therefore don't merely change the form of the old lattices or the density of their meshing but transform habitable space itself, the face of the land, tomorrow deserted perhaps by plant roots and animals niches. At the very moment when we are talking about nothing but networks, there are no more networks – merely this new expanse without any measure of distance, without any physical nodes or crossroads where moving things circulate. How then are we going to inhabit it?

Listen to the echo of the highly moral question: who is my neighbour? Whom should we call neighbour now? Sometimes indifferent to those who pass before my door, or to the door opposite mine on the landing, I often communicate with the other hemisphere, depending on work, speciality, languages learned or friendships formed in the course of my travels; but above all, work now is no longer tied to the length of arms or to the range of action of simple machines but to the information issuing from banks scattered across the Universe but able to be consulted at my home console. There is no longer any here once everything lies here. As Harlequin, the Emperor of the Moon, said long ago in the *commedia dell'arte*: everywhere is the same as here.

The nodes or crossroads on the diverse networks were distributed in a space dominated by their concentrations, of humans, buildings, wealth, information, banks or capital. I even imagine that power, in the sociopolitical sense, could only be defined in a space where such centres radiated. From a given crossroads to the subset of crossroads connected to it by another subset of roads, a kind of star is cut out, a star designating the reference place, the position and strength of this power. Fighting against it opened up two strategies, a maximal one or a minimal one: either do as it does by constructing an equivalent star – whence wars, conflicts, debates, squabbles of all kinds, Athens against Sparta, Rome against Alba, France against Germany, England against its colonies, the United States against the rest of the world or some kinglet against the petty potentate on the opposite riverbank … – or live the reverse and only cut out the smallest possible extension reaching the closest neighbour, perfectly defined amid the networks, in the space of relations and of the habitat: no star, just the minimum of distance to travel; no power, paradise remaining within easy reach. In the new space, the neighbour, today undefined, isn't imposed upon us according to his position but depends on a choice in the possible dispersal. My neighbour lives in Florence or Melbourne. We have just acquired the freedom of choice of our neighbour, a new adoptive dilection. Why?

Ubique: *Anywhere*

Because the old here of the philosophers, or the being-there, more recent, or even the rootedness, is now translated as anywhere. In order to wander once again or to inhabit at leisure, we don't return to the nomadic era since the reigns of the sheep and the horse are ending as are the reigns of wheat or green grass, since livestock farming is being practised in batteries and agriculture hydroponically outside the soil; we also don't return to the era of hunting and gathering, in which we depended on the conditions for flowering and animal lovemaking. We sail, of course, but without budging, from anywhere; this doesn't define classic stability. Here, that is to say, anywhere.

We live without distance from others, of course, but above all we can no longer calculate any interval since we do not know how to assess where to measure it from. When Descartes invented the coordinates of his geometry, the origin of measurement, at the zero point, strongly resembled a node or a crossroads, the foundations dug in the peaceful earth before building the walls of the house, here or my home. The abstract expanse still retains a hole, a sort of pivoting – I hardly dare to say agrarian; I hardly dare to write masonic; I hardly dare to add stay-at-home – root. The formal expanse still refers to this stake, like M. Seguin's goat. When this fixed point disappears, all reference becomes erased. No more link, no more tie, no more attachment, no more servitude.

The crisis of reference began, in habitable space, with the old question: where? From where? *Ubi, unde?* Whereas we used to live in geometry – whose 'meter' said both metric and measure, both our distance to the house and a wisdom whose Egyptian, Greek or Latin names always more or less translated this mensuration – we are now living without metric, as though in a kind of topology. Are we going to build castles in the sky there or Utopian cities? We live less at zero distance from whomever in the world than in a space where distance, incalculable, loses all meaning. A smooth expanse and without any singular hitching point. This is what causes fear. Why?

Who, I?

Because this fixed point, in the old reference, arose from elsewhere, from Mother Earth, from our fathers or from the homeland, from the farm, from the city, from the source of food, from the protective collective … Where are you from? *D'où êtes-vous?* This question strangely identifies the where and the you, the *où* and the *vous*, the place and the I, the being and the there, the me and the alfalfa square it possesses. Without here, no more me. This is enough to make the grumblers take fright: of no longer existing, the poor things, for no longer knowing where they are setting foot. As though the I had to plunge into a space, as though it belonged to a subset it hadn't chosen. The word 'subject' transports inside itself a little of this strangeness since it says of itself that it lies under. As in a hole. As in the zero point of Cartesian measurement. As under a root. As buried: being-there or 'here lies'. As in a grave. As under a funerary slab, crushed by a rock, among the bodies of the ancestors. As in the death that in the past made us into the humans we have become. This tie that was never in doubt seems bizarre to us today. Strange, tragic, oppressive, haggard. A cause of violence and war. Our bloody foundation. No, I am no longer the place where I am, the place where I come from nor the one I am going to, since I don't know it; in other words, I would be anyone. I free myself from this tie, from this attachment, from this being chained. The I is no longer identical to a serf attached to the soil, to a slave shackled for debts, to a convict tied to his iron ball, to a miser welded to his property, to a victim buried in his ground with only his head sticking up, exposed to the flight of vultures. The referenceless I, fixed and mobile at the same time, changing, yes, shifting and diverse, shimmering according to those I encounter in the Universe, no longer depends on anything but me, is only defined by the tautology, flat and white, of identity, never by the slightest tie of belonging. Yes, this old link seems to me to be more than bizarre, wrong, criminal even, since by making belongingness and identity (the epsilon of belonging and the three little bars of identity) equivalent, it is committing a logical and mathematical error, of course, but also and above all a crime against humanity since racism, the root of all hatred, consists precisely in mistaking identity (who you are) for one of your belongingnesses (local origin, family origin, sexual origin, tribal origin, national origin, religious origin or some other). You are only where you come from. No, I am who I am; that is all. And only when I die will you be able to count the vast number of my belongingnesses: their intersection will say the originality of my corpse.

Homo universalis: The end of nostalgia

The universality of the new humanism isn't merely due to the power we have acquired from attaining the global through the energy of the atom or the information of signs; it also isn't due to the all-powerfulness, in the sense of totipotency, recognized in our bodies and our understanding, but also and above all to the expanse we now inhabit. Distanceless space implies a spaceless I. We no longer inhabit geometry, the Earth or measurement, but a topology without metric or distance, a qualitative space.

This qualitative space eases the suffering of trips. That entirely special pain came from the fact that we left those we loved, not only that dear head and those radiant eyes but also places and their familiar things. A seaman's wife waiting on the dock, a Norman sailor wandering on exotic shores and finding his sister in a Chinese brothel. The old humanism was coloured with nostalgia. Not only did we miss, in distance ports, the steeple of our childhood, but every human on the foreign earth mourned for the paradise lost, the heavenly Jerusalem, the valley where milk and honey flowed, the true place of origin and serenity. Homesickness constituted the human. Its existence depended on these unique relations. Proof that place created the I. Not necessarily here, but the garden of my dead mother. Proof as well that space caused sickness. Homesickness shows that the sickness varies with the home. Ontology, nostalgia: the one says in difficult and haughty words what the other one lets everyone understand in songs.

We no longer have space sickness. Where are you from? From anywhere. From the Yangtze, from the Amazon, from the Garonne and the anastomotic beds of the Amur. I flow and don't put down roots. *D'où êtes-vous?* From the Black Sea and the Yellow Sea, from the Red Sea and the seven blue oceans, a mixed-race Harlequin with the thousand hues and white colours of the waters. I sail. Who are you? I fluctuate, percolate and am not. Like everyone, I inhabit the world and its time. The farms and cities, formerly stable, like their pride, descend the rivers all the way to the seas.

PART THREE

THE OTHERS

The Argument
The Event of Communication
Contemporary Humanity
The End of Networks: The Universal House
The Third Loop of Hominescence
The Others and the Death of the Ego

11 THE EVENT OF COMMUNICATION

Without communication, there can be no life

My body seeks to feed itself, a habitat, always, and sometimes a partner. Just as the porous membrane of a cell closes and opens that cell to a thousand exchanges, desired or refused, my skin protects me and is perforated with orifices through which I nourish myself, perceive sounds and light, cold and odours, excrete and copulate. Without this first communication, no living thing would survive: it maintains and characterizes life, from bacteria to the formation of multicellular organisms, from individuals to their reproduction, from niches to the entire biosphere, in which millions of species are brought together in tight groups.

Thus, through exchanges via various channels, living things transmit, receive and store energy, said to be hard, and information, said to be soft because its flows don't surpass the negentropic scale. Without these functions, they would die. Biology, in the broad sense, still lacks a general theory of signals, which would deal with these questions directly, the whole of which would cover it.

Without communication, there can be no society

If communication were limited to preserving life or perpetuating it, my life would be reduced to the above-mentioned functions. When I talk, for example, I don't merely seek to eat, drink or mate. A partly political animal, I have a duty to my family, to my profession, to my city, to a public, to humanity. So I dedicate myself to teaching, to professional relationships; sometimes I even vote. I avoid power relations as much as possible, but the collective, they say, forces one into them, thus adding energy to information. It is understood that society must be organized, managed, administered; is information enough to do this? Must violence and death, must the exchange of hard energies truly accompany this effort, which in principle only requires soft energy? By a feedback loop, as soon as social ties appear, they take charge of certain vital relations so that the second communication sometimes replaces the first one so perfectly it hides it: parasites and leaders, with their voices, command to have dinner or to have something to drink brought to them. Thus the social sciences rarely see the living foundation of communication.

Therefore society is constructed through communication; without the latter the former would perish. When Montesquieu had the forms of government depend on the size of the states, he meant by this that they vary according to the spread of power and information. Writing, the horse, the railroad, the sailing ship, the telegraph or the airplane made and unmade empires, it is said.

Without communication, there can be no self

I'm not merely a set of organs and social roles. Of course, gestures and words do aid organisms and collectives to survive. But, mysteriously, in the unbroken meditation of an interior dialogue and by a hundred friendly conversations whose style elegantly avoids the social and excretory functions, language constructs me.

Living things exchange energy and information; they communicate by hard and soft; the collective might only need information in order to organize itself or persist, but power, conflicts and debates add energy to it, perhaps useless energy, violence that, in the absence of language, neither plants nor animals can do without; when the hard comes to this soft, it leads humans towards the world's misfortune. Lastly, in the third case, energy yields to information, but what passes, then, between two interlocutors becomes just as light in relation to information as information is in relation to energy; the soft of the soft alone remains, the way we say the best of the best. This third information flies, so airy, sylphlike, agile, lively and luminous that people wiser than I called it spiritual: Genesis's *ruach*, whose breath hovers over the primal waters, the Christians' Holy Spirit with its seven gifts or that Greek wind that is so subtle it produced the Latin *anima* or soul. No one has any need for social power or glory; everyone seeks to feed themselves, not only with bread but also with words.

Let's call this last communication communion, a communication that's rare, improbable but as vital for the self as the first two are for the body and the group. No one is born before an other says to him or her: I love you. No one survives lack of love. It gives existence and saves.

Who?

Talk about communication, in general, about television or radio, even language or the internet, in particular, remains abstract if one does not ask the question: who, individual or collective, communicates? Answers such as the living thing, which would die without this, the mind, which makes it its food …, still remain just as broad and abstract since they concern all things. Observe who is contented or not contented with silence and obscurity, who benefits from both behaviours. Communication increases with power and power with messages, sometimes, noise, often, but above all interceptions. As with living things, many social phenomena can be explained by simple parasitical operations, by diversions of the messaging system in favour of the interceptor.

All communication gives publicity to the person broadcasting it. When it announces itself to be such, this publicity seems more honest and truthful to me, by this very metalanguage, than the broadcast that devotes itself to advertising while concealing that it is publicity. Additionally, under the word 'publicity', a powerful meaning is concealed, the essence of the public. To whom is one communicating? To a public. The narrower it becomes, the better one approaches the me and the you, what I have called the soft of the soft. The more it grows, the more power strengthens and parasites multiply. So technologies intervene: Stentor, yodelling, megaphones or drums only affect squads, villages or valleys; striking coins already allowed the king to spread the image of his face, in profile or full-face, and to get himself known by those, numerous, who desired the gold coin or those, rarer, who received it; fires lit on the island of Pantelleria by the Carthaginian armies transmitted messages from Africa all the way to Sicily and Southern Italy; from Chappe's semaphore telegraph to the internet, the public involved grew, at least in number. So abstract concepts, public opinion and the general will, for example, enter into practice: the new technologies employ them. Towards the end of May 1968, few policemen surrounded the

Sorbonne or the Odéon, or even the National Assembly, but the government sent regiments to forbid access to the Maison de la Radio. Finding myself there, I was delighted that my society's decision-makers knew, in practice at least, that they were already bowing under the reign of Hermes. But who precisely bears the name of this god? Who holds these technologies on a daily basis? And who, being in possession of them, excludes the others from this post? Nothing could be easier than to answer this question.

Answer

If we divide the information communicated in the form of language into its aura of seduction, its performative character and its truth value, the media, for the first one, the administration and the judiciary, on second account, and the sciences lastly hold its three crenels. Aren't we stating here the three sovereign powers of Western collectives, insofar as they don't meet anywhere with counterpowers? Whoever wants to criticize the media has to pass through them; whoever opposes law or science either is wrong or will receive the police at his residence. Who can change your life? The journalist, the administrator, the scientist.

But, since seduction always prevails over law and truth, let's give priority to the journalist: after having taken the floor away from politicians, but without any test by election; expertise away from scientists, but without any knowledge; education away from parents and teachers, but without any professional ethics; prosecutions away from prosecutors, defences away from lawyers and verdicts away from judges, but without any judicial skill; questions away from the police by forcing everyone to answer, but without having to answer any questions themselves; after having seized the place of every decider, but without any responsibility, in short, every glory without any obligation or any sanction and every power without ever paying for them with a counterpower, the journalist lastly steals, while hunting down every breach of morality, confession from priests, but without any secrecy or pardon. When the parasite or universal interceptor takes every place like this, how, through this omnipresence, wouldn't he invent the entire social real?

What?

The same abstraction without analysis clarifies the transmitters, channels, memories and receptors of said information. The messenger of

Marathon, d'Artagnan on horseback, the steam-powered Human Beast, the transatlantic liners of the Cunard Line, and Airbuses do not carry the same messages, along the same paths, with the same speed and towards the same number of people as radio, television, the internet and its browsers.[1] They therefore do not, whatever may be said, bring about the same society, even if society contributes towards constructing them. A causal feedback loop makes them into two causes and two results. Disregarding the various technologies amounts to forgetting these considerable differences but particularly amounts to ignoring that objects are a peculiarity of humans (the other living things don't have objects), to ignoring therefore that the collective in large part depends on its objective technologies, which depend, in their turn, at least today, on the sciences said to be hard, even if society contributes towards these sciences. So in the end, the soft sciences depend on the hard ones. The road doesn't only carry the queen's messages but especially the carts of rocks without which its pavement would be pitted with potholes.

In the collectives I have had experience with for almost a century, I don't know of any notable change that didn't shoot up one day from this hard source. The new body was born from sulphonamides and analgesics; the liberation of women arose from the pill; agri-food came from biochemistry and modern communication from electrons and optical fibre. Does the West owe its long peace, so rare in time and space, to theorems from before the nuclear terror? The demographic explosion, as well, depended on physics and pharmacy. Cite a single ethical problem that didn't, in its turn, come into being in these sciences. No political decision could have, at the same time, had comparable effects. Such a decision would have put the brakes on them instead.

If the men and women who, over the course of the century, had transformed society must be called 'great intellectuals', then cite the women and men who, in the laboratories, had worked at chemistry, agronomy or electronics instead of philosophers, who were so little 'engaged' in a time whose factors of change they were ignorant of. Honestly, I don't see what Jean-Paul Sartre, Malraux or their regressive epigones had changed in our conditions of life, whereas we have all lived differently if not better since Fleming for the comfort of our body and since Shannon or Turing for our relations between ourselves. The sciences of nature change nature, body and world, and even, sometimes, cultures, my own and the cultures of others, whereas the knowledge of societies describes them without transforming them. This is why this book often digs beneath History before rejoining it, sometimes.

How?

How does the listener receive the transmitted message? How does this latter circulate in the selected channel? Who intercepts it? We have already answered this question. Two ideals, highly different, of communication are opposed to each other: perfect reception and the absence of filtering. The political ideal of equality implies a requirement for messages clear enough that everyone can understand them. Each person has the right to understand the world, his or her body, the Universe and the others. I have fought under this flag since my childhood, for political equality in fact depends on it.

But communication, such as professionals practise it, confuses this equality in reception with a maximal filtering. 'We are giving you one minute and ten seconds to explain to the general public the reverse transcription of retroviruses: be clear and above all concrete.' This is an example of a sentence I have heard for several decades. No one can bring off such tours de force without lying, not popularizing but misleading about the merchandise. Equality doesn't imply a ban on telling the truth. This poses a question that's impossible to avoid.

Generalization of Aesop's theorem

If we call tools that function at the entropic scale 'hard technological' and machines that allow the exchange of information 'soft technological', what do the soft technologies add to and what do they subtract from the three communications I just defined, for the body and life, in the collective, for personal salvation and cultural delights? Do they transform them?

Yes and no. Older than writing by its date, Aesop's ancient formulation said that a tongue was the best and the worst of things. For, through it, I can feed or starve you, sing your praises or slander you, thus perching you on the summit of the Capitoline Hill or throwing you to the bottom of the Tarpeian Rock, tell you I love you or that I hate you. This duplicity of the first communication route, since it travels through the organ of exchange par excellence, this tongue that eats, drinks, talks and commands, generalizes to every means of communication, which can primarily be considered to be routes that are independent of their content or neutral in relation to the messages they transport. Thus, the telephone has saved lives, but it can also kill; in any case, it helps but sometimes transforms into a nightmare. Either no one travels along this freeway and its route, always clear, remains the best, or it immediately clogs up and becomes the worst. We can learn anything on the internet, including sciences and medicine, Nazism and pornography. No soft technology escapes this double logic, on this side of good and evil, of the false and the true, of life and death.

Of course, the ambiguity of this generalized theorem by Aesop also applies in the absence of soft technology: I became a passable algebraist because my tensorial calculus professor held me in esteem, a priceless treasure I gave back to him with profit, but a sorry speaker of living languages for the converse reason that my English teacher hated me, an ignominy I returned to her with interest. Face-to-face campus-based teaching falls under the same duality as the one that burdens the soft technologies; let's not accuse machines of situations established before and without them. Let's avoid the temptation to consider the 'natural' to be good and the artificial to be bad: here, failure and success are shared as much by living presence as they are by distance learning and seesaw both of the latter with their doubt as well.

This double logic lastly explains the disagreements that incessantly set in opposition the people who judge the soft technologies along with their relation to living communication, whether social or cultural. So before judging or deciding what seems good or bad, before setting off for battle, let's observe, explain and understand. The battle Don Quixote did against windmills already staged the man of the book, Heidegger-style, raging against high technology. For at that time, these windmills, the ne plus ultra of practical modernity, disfigured, with their unfolded blades, the agrarian landscapes and pastoral meadows in which the chivalric romances unfolded their virtual marvels.

Even though every road remains indifferent to what transits along it, can the same theorem be extended to transfers of hard energy? Certainly, since tools aid and weapons kill. In any case, the living thing is equipped with an immune system or various filtering strategies the way the collective is protected by borders and customs checks. Since we can die from a lie, I hate, with all my soul, scheming and cheats, more cunning than simple gate guards; but I have a passion for parasites and their double game of interception.

The parasite

The same antinomy applies in fact to the principal interceptor of all communication, whether 'natural' or technological, the parasite. A third intervening in every relation, it procures a habitat, food and often even reproduction for itself in and by a host, sometimes precipitating it towards death; thus the one who intercepts information grows fat and wealthy; but conversely, disrupting habits, it can make the couple it forms with the host bifurcate on to new paths of life by forcing it to invent a symbiosis. Specialists find examples of this almost every day. If we had, in biology, a good general theory of signals and communication, it would appear as an evident fact that there can be no life without communication, of course, something we all know, but also, in a refined and deep way, that there can be no living thing, whether single-celled or multicellular, no species, no reproduction, no lineages, no development or evolution … without a multiplicity, always decisive, of parasitical operations. Beyond the cuts, sometimes lethal, it brings about in every communication channel, it sometimes re-establishes others and thus promotes the new, whether organization or evolution. The parasite holds life and death, the origin and the end, exchange and gift, time and composition, good and evil, the false and the true, order and disorder. So practise seeking a power or some dominant one who doesn't behave parasitically.

Soft technological intervention on three fronts

Through language as on the telephone, on the internet as by gestures, I can procure, in an equivalent manner, the address of the baker, understood as a supplier of bread and social function, and the address of my beloved. In other words, all soft technologies intervene in vital, collective and cultural communications. But, conversely, the mail lets us read a writing, and the telephone allows us to hear a voice; this changes everything. Letters can lie, but a too assured or trembling intonation confesses, without saying them, fatigue or indifference. How many times have I noticed, on television, the completely objective honesty of the portrait in motion? A son of a bitch hides his ignominy badly; the killer overexposes his hardness; honesty rarely deceives. Here is, on this point at least, a truthful medium! No soft technology is equivalent to it.

Do friendly dialogues and subtle loves, whose delights contribute to constructing everyone's identity, escape soft technologies? In other words, do these technologies kill our best communions? Whoever never liked stars, song, the cinema or football ought additionally to marvel at the vocations created by images. We remain virtual in three-quarters of our acts. Why wouldn't love pass over telephone wires? The telephone's success even seems to me to come from the fact that it favours its pipe organs and its delights at least as much as the Stock Market and business, cheating and threats.

So soft technologies become adjuvants and obstacles in three types of communication, each with its style, its advantages and its disadvantages. These technologies transform them, for better and worse.

The extrema

I have already explored this logic (*Zola, feux et signaux de brume* [Zola, lights and fog signals], pp. 295–303). Better and worse than everything, God Itself and a thousand devils, savage exploitation and angelic charity, death and life, destruction and construction, precious and excremental, real and sign, being and nothingness …, who doesn't appreciate and despise money? Occupying the global interval between these *extrema*, it invades ethics and the culpable, the religious and the atheistic, violence and peace, the collective and the individual, politics and the living, theory and practice, the stock and the flows, chance and causality, the solid and the liquid to the point of the volatile, the necessary and the virtual, ontology and the cognitive … in sum, every possible spectrum. The set of these avatars gives it the status of general equivalent. Where it reigns, there can be no other values since it replaces them all. Odourless, money whitens everything it touches as though it summed up every colour.

With communication, have we touched upon a new general equivalent? Since the paths, whether 'natural' or technological, the flows, whether energetic or informational, the nodes, most often parasitical, obey this multivalent logic, the fabric they form tends towards a global equivalence, finally attained when a valid unit of account on the internet will have replaced every world currency. And already, this or that good is worth more or less according to the information you can obtain about it: everyone has his account in the data bank. Will money take on second rank, withdrawing before an equivalent whiter and more general than it is? This is why we insist on the terms 'information' (multivalent because stripped of meaning and able to take on any meaning) or 'communication' (so general it can mean or say anything).

The white potentiality of the virtual

What roles, for example, do media networks play in today's societies? The already evoked answer: all of them. Political roles, yes; judicial roles, more and more; ideological and police roles, assuredly; educational roles, in overabundance; the role of distraction: this is their origin; moral roles: daily ... No, it doesn't have to do with a fourth power, in Montesquieu's sense, but with a general equivalent, with the potential for every power. With totipotency. This general equivalent even seeks to destroy those powers that, exceptionally, it cannot exercise: the power of the sciences, for example, and the power of religions, the closest to it, in truth, since religions also transmit truths about the real as such. Through their internal dynamic, these networks therefore construct another real, another society, a new ideology, another education, another politics and so on; in the end, another mode of being and of truth, another kind of money. Giving up the role of representing a real that's already there, these networks take on the role of creating their own real. Not necessarily through their status or their technological qualities but through the general equivalence of communication. There are only two common places left where every person finds himself, without any distinction, and where everything, without distinction, is of equal worth: money and the media, masters of the *extrema*. At this moment, the battle is raging, and alliances between the media and money, by which each can take the other from behind, are multiplying. As in life, the parasites hold everything.

The cognitive: The question of truth

I remember my youth, in which I abandoned mathematics one day so as to devote myself to belles-lettres, or again, the hard for the soft; the experience of this passage from the sciences of the object to the narratives of subjects still remains alive in me; on the one side, every morning I had to give myself over to violent efforts to understand; on the other, I understood everything easily and immediately. At least, I thought so. This distance dazzled me then and still does now. The social sciences didn't require great effort in order to understand that the rich are distinguished from the poor in thirty ways or that scientific discoveries presuppose money and power. But who has no difficulty in grasping superstring theory? This distinction would have no great importance if these truths, more or less accessible, had the same impact on the evolution of societies. But they don't, and by far, have the same effect.

The egalitarian space of communication resembles the space of the street today, where no one any longer distinguishes, the way they did fifty years ago, the priest from the military man and the worker from the student.

I'm not complaining about it, quite the contrary, but everything has its price. For this homogeneous public space, in its turn, copies the space of the narrative or of the social sciences, in which the true is confused with the immediately accessible. No. Certain truths twist common sense, the truths of God and its ubiquity as well as those of quantum mechanics and its probabilities of presence. Yes, both spaces, the one that tends towards the democratic ideal and the one constructed by the media, praiseworthy assuredly for having carried out their encounter, nevertheless copy this state of comprehension proper to narratives or to the social sciences, which never leave ordinary intuitions. Yet through an effect I am explaining poorly but which all the same never ceases to take place, these social truths, which are so quickly understood, have only little impact on society itself, whereas the most difficult truths, concerning atomic particles, for example, or reverse transcriptase, unleash terrors or curings, which in their turn bring about social quakes that make time bifurcate. A paradox: society changes with non-social truths; why do they remain hidden? Subjects change with the truths of objects, truths so little accessible.

Contrary to what is supposed by those who hold communication, that distance, residual and hard, that they try to erase in the name of the public is seen, understood and respected by the public itself. For the people, to which I belong, live and think more archaically, more anciently, more rooted in objects than the elites; these latter believe that only subjects, both individual and collective, exist around them and that everything is settled by power relations. They err by too much 'acosmism', a strange belief that society dispenses with the world and lives outside it. 'Everything is political', this is a saying from the dominant because, for the decision-makers, everything depends on the power deployed in the relations of dominance. The dominated, for their part, excellent experts in the functioning of societies for seeing this functioning, transparently, from below, have known, since the foundation of the world, that the necessity of objects exists first, a necessity stronger, through its constraints of impossibility, than the possible caprices and contingent demands of subjects, whether individual or collective; they even know, obscurely, that society only exists because objects exist. Objects alone gather us together. The dominant live in the mode and the dominated in the modalities. This problem of Truth, which gets lost in communication, is today becoming more decisive than it was believed to be for the future of our society since History, whether subjective or collective, bifurcates according to the truth of objective mutations.

Examples touching on questions as important as energy and health: do we need to be afraid of the waste from nuclear reactors or the expansion of GMOs? If yes, let's close the laboratories and power plants. The prudent and often optimistic specialists are opposed by fears fuelled by the news of deaths and diseases, spreading terror and pity. What should we protect ourselves from? What should we decide? Individuals and public opinion find themselves in the presence of several interlocutors, personal and collective in their turn: scientists or Science, businesses or Engineering, capitalists or Finance, journalists or the Media, administrators or the Law, the representatives of the people and the decision-makers or Politics. In person or globally, together they form a metastable network, buzzing with communication, now partly objectivized on the internet, where messages and noise circulate. What or who should we trust amid this fluid vortex of contradictory information? What does Truth become there? Contrary to appearances, debate about it is less in evidence than the fuelling of a great fear in the face of mythic objects that have become terrifying phantoms. But who finds their advantage in this increase of anxiety? Is a new age of darkness coming?

The archaeology of Truth

This question of Truth can't be settled without our remembering its ancient history. Here it is. Most of our ancestors died without leaving any trace of their passage. Even those to whom we owe the most gratitude, the first farmers of corn, oxen and sheep, the inventors of bread and wine, of the forge and the alphabet, were plunged into a loss without return by death. We don't know their names. They have, the ancient Greeks said, crossed, in Hades, the river Lethe, whose name signifies a definitive forgetting. Once having passed through this semi-conductor transit, no one retraces his steps. Except for a few rare exceptions, two or three per generation: Achilles, whose anger shook the Trojan walls, Ulysses, whose voyages explored unknown islands, remain in our memories because they recrossed, in the other direction, the infernal river. Why? Because, by chance, a poet of genius, Homer (did he even exist?) sang of their exploits in the *Iliad* and the *Odyssey*, whose verses all the youth of yesteryear knew by heart. In saying *alétheia*, the same Greeks reversed or denied the name Lethe – the *a* that precedes it is called 'privative' – in order to express with exactitude the return of the two heroes, who were made into ghosts, come back to haunt the Earth by their glory; so *alétheia* describes the voyage, so rarely reversible, across the river, and the memory rediscovered beyond forgetfulness. Yet today we translate *alétheia* as truth, even though it only used to express that renown sung of by Homer, whose narratives staged the sailor's ruses and the warrior's courage. Truth therefore wasn't distinguished from the glory acquired in the best communication routes known to that archaic era, the epic poetry spread in the schools and public squares.

Today we imagine poorly the efforts of the first astronomers, of the Presocratic physicists, of those who proved the theorems of geometry or arithmetic, to replace with faithfulness to the observed facts or with the necessity of precise reasoning that social glory whose halo rendered men immortal, men who had perhaps never battled or sailed since it sufficed for a writer to claim it for everyone to believe it. We imagine poorly the efforts of the scientists and philosophers to move truth into the house it subsequently inhabited; nearly all of them appeared before tribunals, the only deciding authorities, in fact, to already require that facts be established. Almost all of them heard themselves be sentenced and paid with their lives for the invention of truth, in the modern sense of the word. The slow and difficult construction of proof took place before the court. The new truth was set in opposition there to the ancient *alétheia*, that is to say, to collective glory, to the public's belief in the exploits of this person or that person, exploits sustained by dramatic facts, dense with sensationalism, terror and pity. Since glory always wins out, we unreservedly admire this rare miracle, yes, this true Greek miracle, in the course of which truth, in the contemporary sense of the sciences, won out, in a short space of time, over the communication that related, at that time as it does today, the tragic horrors – massacres, pillagings, rapes, fires – of the Trojan War and the magic narratives of the distant voyages in which Ulysses, a reporter, discovered strange landscapes where monsters grazed. Who admires this unparalleled exploit at its precise value: the victory of truth over glory?

The theology of Truth

A similar history unfolded on the side of religion and shook the Fertile Crescent just as violently. Polytheism, a set of social cults, fabricates its gods by means of human sacrifice; the light of their glory feeds its brilliance with the fires lit to slit the throats of the victims and with the horror of these murders. A thunderbolt rends this bloody and repetitive history: monotheism considers these archaic gods to be false because, it says with reason, man, by killing man for glory, fabricates them in his image and to his likeness. So reversing polytheism, it turns a single and absent God into the producer of man, the world and the Truth, never the reverse. As a result, this truth is born here from the just criticism of those gods monotheism is right to call false gods. To the one God then, transcendent and inaccessible, and to Him alone, goes all honour and all glory.

Glory: What should we call contemporary?

Therefore truth was everywhere and always set in opposition to glory, but glory inevitably wins out, except in these two evident exceptions, which we inherited by chance. The same opposition continues, but if we forget it, we risk losing this inheritance, one of the most precious. Yes, on this side of monotheism and Greek science, we always risk returning to the era when no one knew Truth, in the modern sense. In Homer's stead, a multiple poet to whom a hundred ancient Greek lyric poets lent their voices, lending them as well to the numerous gods born of serial murders, today hundreds of thousands of messengers communicate deaths, wars and tragedies, making this universal glory shine again, a glory whose noise largely covers over the truth. Why, today? Because, nothing having changed in this regard, everyone vies with each other to get his voice heard, even religions, even the Nobel Prizeworthy, for the sake of a similar glory.

This glory forms the social bond. It constructs public space, whose transparent volume adds up politics, the media, the judiciary and ethics, social sciences and social practices mixed together. Its cement makes democracies, the new kingdoms of this world, hold together. Subtle, it has taken the place of the single God, going against the precept, all in all fairly prudent, of giving glory to Him alone in order to preserve ourselves from quarrels. Glory resonates as the high note unifying and dominating information. Neither the content nor the form of the discourses nor the music, neither the noise of the combats nor their victims count; only the deafening peal of its white tonality matters.

Once again, what should we call contemporary? The society of communication is built on glory just like the archaic societies during the Homeric or polytheistic eras. No voice prevails over its monotonous tone, whose harmony, in the acoustic sense, seems to bring about harmony in the social sense. The differentiated, shimmering, mosaic, harlequin, hesitant, feeble (in total) lights of truth disappear beneath its wax polish.

Doubt

If a person wanted to stage a contemporary doubt as effective at the one Descartes used when he was seeking the truth, the Evil Demon whose spell shook the philosopher would have to resemble the Evil Demon that tempted Jesus in the desert when his fast ended. 'Showing him all the kingdoms of the world and all the glory that goes with them, he says to Him: "I will give you all these things if, bowing down before me, you worship me'" (Matthew 4:8-9). The sciences, philosophy and religion will only return to being sources of Truth, of its patient construction and its patient propagation, on condition of renouncing glory. They have to set out for the desert again.

For at least forty days. For we will never be able to completely rid ourselves of this glory, which sticks to us for two reasons, global and local: its single note now fills the physical, technological, political and social, even democratic, universe; each of us has need of it, at least in small doses. Wisdom patiently negotiates the vague encouragements glory gives and the serious traps it sets. Once again, it resembles money, without which no life can survive, no project can be built, but which sows death around itself.

Truth is born, a little and sometimes, from colloquia, meetings or debate, which power and glory on the contrary more often depend on. The new is rarely born from communication in general, which tends from all sides towards monotonous vitrification. Proportional to rarity, information, on the contrary, the true kind, shy and hidden, appears more frequently to solitaries and the silent. Difference shoots up from singularities; but singularities would disappear without a global, single and transparent backdrop. Little glory favours truth a little, but a lot dissolves it into totality.

12 CONTEMPORARY HUMANITY

New technologies

Our body listens, cries out and remembers. Bacteria, algae, mushrooms, plants and animals likewise signal their presence and perceive the environment, each in their way; without these exchanges of energy, of course, but of information as well, no organism would survive. Before becoming human, communication characterizes the living thing as an open system; cells communicate with each other in bodies as much as bodies with each other amid their ecological niche. At the small scale (chemical reactions) and at large scales (thunderstorms and galaxies), energy and information are again exchanged within inert matter.

We humans added to these purely physiological or physical performances a wide variety of man-made objects intended to take over the reins from our bodies in its communication activities; this arsenal of messaging systems and semaphores varied over the course of time. Quite recently, electronic technologies once again completely changed the set of tools allowing the reception, storage or preservation, broadcast or transmission of information.

This recent change concerns time, space and the relations between humans.

Time

We have already experienced at least two complete changes of the same type: the invention of writing and the invention of printing. Engraved in stone, bronze or wax tablets, before it could be read on papyrus or paper, writing contributed, in a decisive way, to creating the first cities, in the Fertile Crescent, to creating large states organized under the rules of a written law (the Code of Hammurabi, Mosaic law), accelerated commercial exchanges and made them more flexible thanks to the striking of currency, gave their rapid development to the abstract sciences and pedagogy, in ancient Greece, as well as to the monotheistic religions, which can be defined, rightly, as cults of Writing. Better yet, today we divide human time into two distinct parts, prehistory and History, the latter beginning precisely from the appearance of the first engraved texts. The large political, religious, economic and scientific stabilities that have come all the way up to us therefore proceed from tools peculiar to processing information, which don't so much vary in History, as I just said, as they command History, since writing caused it to be born.

As soon as, with the Renaissance, printing appeared (the first assembly line or mass production), the Italian banks transformed commercial exchanges in the Mediterranean, in which bills of exchange replaced currency; they consequently launched the first capitalism; the circulation of books favoured the individual independence extolled by the Protestant Reformation, therefore political democracy and civil law; storing them in libraries lowered the value of doxographies and, by lightening the memory, put the observer in the face of raw facts and thus contributed to the appearance of mechanical and physical experimentation; in sum, printing engendered modern science; lastly Montaigne, Erasmus, Rabelais … drew unusual conceptions about pedagogy from these innovations. The two transformations present a similar profile.

The different impacts of hard technologies and soft technologies

The changes in the recording media for information – 'soft' technologies, at the negentropic scale – therefore seem, by their flexibility, speed and capacity for expansion, to influence individual behaviour and social organization more strongly than the aforementioned revolutions engendered by the 'hard' technologies, at the entropic scale, such as the Industrial Revolution. However much mechanics and thermodynamics have long introduced us to a precise and developed knowledge of the hard technologies and their laws, energy constants or engine outputs, we are still largely ignorant of the laws of the soft technologies, so distinct in orders of magnitude and applications. In the French language, I have therefore kept the Anglicism *technologie* for the set of man-made objects that manipulate signs, that is to say, *logos*, and oppose it to *techniques*, whose energy field of action differs from the first one by a factor of 10^{16}.[1]

A new example: in recent decades, philosophers with delicate hands had us learn in books the decisive importance of hard technologies outside of books, the technologies of mines, factories or workshops; de facto idealists, even the so-called materialists didn't consider the transparent pages they wrote to be soft technologies. Stemming from the necessary actions of living things, and no doubt, more distantly, from exchanges in the inert, the various ways of accumulating or exchanging information govern changes that are less visible but longer range than the changes the high energies seem to determine. My generation witnessed the disaster: the steel, coal and blast furnaces of recent times, on which my fathers believed they were building Europe, quickly joined the windmills and spinning wheels of yesteryear on the scrap heap, whereas the computer is multiplying printers and making the ancient engraving of signs triumph. Far from killing the preceding ones, the invention of a recording medium renews and spreads them. It would have been better to have constructed the Community starting from teaching!

If this rectification of our view of History has any meaning, and, something easier to verify, if the new technologies innovate greatly in comparison to the previous ones, we must therefore expect complete changes and even ruptures of a breadth at least equivalent to those that shook those two events of the past.

In fact, the economy is transforming before our eyes by being spread on the internet, rendering currency volatile before demanding a single unit of account; the sciences have already changed paradigms under the computer's influence; urban and rural spaces are fairly quickly being redistributed, and, faced with all the other ones, each religion is entering into crisis; some of us are seeking a new law since the internet and Science today show numerous lawless places, and everyone is lamenting that politics is falling into desuetude, while waiting for direct elections. Advanced societies as well as developing ones are planning, lastly, distance learning, in favour of a youth who, once more, don't understand the resentment their elders have towards the new culture they don't accept: you might think you were hearing Socrates refusing to write and repeating his praise for oral transmission or the entirely Latin-besotted professors of the medieval Sorbonne in the face of Rabelai's laughter. The profile of the old transformations is reproduced in large strokes.

Space

Let's summarize, again and in a word, the old world: we lived in a space dominated by concentration. A city gathers families, and a street, a quarter or a marketplace groups homes or professions; a business combines means of production and means of communication; a farm amasses seeds and mates animals; a bank, a library, a museum bring together fortunes, books and works; a campus aggregates laboratories, dormitories and classrooms; a lecture hall packs together students …, a book strings together thousands of words, and the concept of a circle pins together an infinity of rings … It has taken me years to understand that comprehension itself, intellection, cognition, in short thought, obeyed, as singular instances, this broad gesture of accumulation, one that's indissolubly material, energetic and informational, alimentary and vital, demographic, collective and social, practical and financial, political and scientific, mnemonic and cognitive. In the self, a few ideas are seated; in the idea, a multiplicity of instances; in the book, millions of signs; in the library, thousands of books; in the city, libraries; in space lastly, cities, farms and roads. Stocking precedes and conditions exchange. Did it start with the grain granaries of Mesopotamia or game reserves like the ones the Australian Aborigines have preserved for tens of thousands of years? Is it a question of a condition for the purely energetic hard technologies?

So ever since we have been human, we have lived in a space polarized around concentration sites, houses, villages and diverse treasures, in particular the very place where I live and to which my address refers: and since roads of all types linked them, we lived in a network space since constructing formed it, inhabiting consolidated it and thinking consisted in reproducing it. The species stocks, the individual thinks – it's the same process. We couldn't have survived without these concentrations and these circulations that conditioned life, the individual, the collective, practices and theory; tirelessly, we never stopped inventing new networks.

And now computers are completing this segment of hominization. For if these machines can be called universal, they merit this title precisely under the heading of concentration. What need do we have of collecting books, signs, fortunes, students, houses or professions since the computer has always already done this? This general problem of stocking, which we sought to resolve and which we have worked frantically on ever since our own emergence, has found its solution, one that's not only real, but also virtual: every question of this type finds multiple possible answers depending on its conditions and constraints.

The new technologies render the current concentration obsolete, I mean some pile here and now. The speed of communication concentrates virtually everywhere, ad libitum, all or part of what's availably connected. Unlike the old ones, the new machines replace the function of preservation with fast transmissions. The set of roads is sufficient for synthesis. We no longer stock things, but rather relations. Exchange relativizes stocking. Do we need to rethink capitalism?

I laughed for several years watching the Grande Bibliothèque go up on the quays of the Seine at the very moment the accumulation it was achieving was becoming useless and absurd. The Pharaonic decision-makers who, without regard for the poor, squandered so much money on its four towers evoke the maharaja who, in the seventeenth century, erected, in Delhi, gigantic sundials in order to observe the sky with the best precision at a time when Galileo was pointing his telescope towards Jupiter and discovered satellites there. But how can we not excuse these Hindu princes for their backwardness since thousands of kilometres separated them from Renaissance Florence? The Elysée, assuredly deaf, was, like everyone, as close as can be to the noisiest global communication, now ubiquitous. So a poorly advised president bestowed four sundials upon Paris when anyone at any address at any hour could reach a connected spot where there would only be a single book. What good is piling them up?

This ancestral gesture brings the preceding decades back to a second prehistory, one in which memory needed sites, the wealth of warehouses and clever men. When routes multiply, cities diminish; the more neurons there are, the fewer engrams; relations increase to the detriment of what the Middle Ages called substance.

Concentration yields to distribution. As soon as we have every possible access to goods or persons at our disposal on our portable consoles or mobile phones, we have less need for express constellations. Why have lecture halls, classes, meetings and colloquia in a given site, why even have a head office, once courses and conversations can be said at a distance? These examples culminate in the example of the address: over the course of time, it referred to a place, one of residence or of work; today, email addresses or cell phone numbers no longer designate a given spot: a code or a number, pure and simple, is sufficient. When every point in the world has a kind of equivalence, the couple here–now enters into crisis. When Heidegger, the most widely read philosopher in the world today, calls human existence dasein or being-there, he is designating a mode of inhabiting or of thinking on the path to extinction. The theological notion of omnipresence – the divine capacity to be everywhere – describes our possibilities better than this funereal 'here lies'.

A digression concerning addresses

Ever since the genius wife of a hunter ancestor dressed in skins cultivated a square of barley for the first time, their agrarian couple settled around a dwelling they erected next to this treasury of abundant food. Thus animals never stray very far from their niche, nest, area, den or hiding place.

What is an address? A thing and a word, both endowed with precise roots. An exact announcement of this habitat, an address therefore gave the correct direction towards it or prescribed on a message its destination towards this residence. Yet, this term, in addition, comes from the king, who, reigning here, defined the local borders of his power this way; from the king therefore and from that law that only changes, it is said, when crossing cols in the mountains. Clever, the mail carrier, every messenger, but the policeman and the judge as well, could summon you in the name of the king and the law as soon as you admitted your rural or urban address. Even the *Rules for the Direction of the Mind* presupposes, as I will claim later on, a space equipped with orientation and direction, therefore constellated with places marked out by geometry. Habitat, power or jurisdiction, rectitude in method and thought, address announced the wealth of place by listing off its characteristics. Even nomads, driving herds in front of themselves, knew how to recognize the tent, the tepee, their movable home grounds. Even email addresses refer to a site occupied by a device that's heavy enough not to be transported easily.

For the first time, the portable cell phone and laptop have liberated addresses from places.[2] I no longer call you only at home, in your office or in the middle of that old square of alfalfa, but wherever you may roam, on the sea, on the summit of the Matterhorn, on the train or the plane, just a hop, skip and a jump away or on the other side of the longitudes. You answer me without knowing where I'm questioning you from, and I listen to you unaware of where the answer is coming from, except that a number tells the placeless source of the broadcast. We talk from code to code: the geometry or topology of the expanse gives way to an arithmetic or a cryptography of numbers. Absent from the local, we find ourselves present in global space. Strollers, wandering or lost?

If the address carries with it this semantic network of correction, uprightness and direction, then when it vanishes, the force of every rule becomes erased. Some people want to regulate the internet: since addresses imply the king and law, they are afraid of losing all law at the same time as places and their destinations. I prefer to dream about the old story of Robin Hood, wandering in the coppices and high forests of Sherwood, where decent folk would lose their way, fearing exposing themselves to the attacks of his henchmen. A lawless place,[3] like the internet itself, where each of us browses at leisure and without any surveillance, such a woods in fact served as a refuge for outcasts, for pirates, in a word, for outlaws. Incapable themselves of surviving long term in their own disorder, they ended up giving themselves a leader, whose contradictory name evokes both this lawless place, the forest itself, and the man dressed like a jurist or a magistrate, called robin by his time because of this garb: Robin of the Woods.[4] This is an allegory of a law stemming from the very place where no law ever took place; this hero's name evokes the judge's robe. Who notices this oxymoronic name? What can a magistrate do among those living on the margins, a man of law among outlaws? Robin of the Woods: this is the coded secret of a story that's truer than history and which tells without saying so of the invention of a new law in a place that was formerly lawless. How did a rule come about? Why did it impose itself even upon people who devoted themselves daily to savage acts of violence? Here's how: they elected a leader. This green-tunicked tale swiftly recounts the process that leads from violence to contract, the very one Hobbes, Rousseau and their successors reinvented by means of wooden cant. Livy's story wasn't recounting anything different regarding the forests in which Remus and Romulus, the future kings of Rome, were born of a shewolf-whore who was a companion to bandits, riffraff and crooks. A new law is always born in a lawless place; it cannot be imported from elsewhere since it has to resolve an entirely new question: how then are we to travel straight in an impenetrable expanse?

Non-place

How are we to characterize this new space? Since the concentrations and their needs, the distances and their constraints are nullified there, the ancestral network, a form that's always repeated and rebuilt, always woven with centres and roads, vanishes, better yet, becomes erased at the very moment when we are no longer talking about anything but networks. We now no longer inhabit this latticed form but rather a qualitative space without reference (whether punctual or polar) or distance. Are we therefore returning to the forest of origins? So we must rethink space, the habitat, the here and the now, the collected objects, the collective subjects ..., that is, philosophy in its entirety, law, and cognition in particular. Going back through the origin once again, we are reliving and revisiting it today, when those who want to legislate the internet are searching for ways to succeed at this. But, in these virtual places, who today can define a subject of law, the objects of a crime, the spatio-temporal conditions for the application of the enacted rules, lastly the public to which this law would be addressed? No address, no borders: non-place. The internet unfurls, for the time being, a lawless place; can we formulate a law for a non-place?

Supposing, at the limits of the impossible, that this legislation became reality, then it would have to keep watch on our gestures, sentences and intentions in real time to such a degree of sharpness and exactitude that the most intolerable law enforcements of History would seem playful leisure activities in comparison to such subjections. Thus actual law remains inaccessible on the virtual internet and, if it comes there, it will kill all freedom. As Robin of the Woods teaches us, sources for unforeseeable innovations sometimes lie in lawless places. With time, new contracts and then a new law will probably emerge from the messages, exchanges and common practices on the internet; the new laws will therefore depend on the internet itself, the way, since the Code of Hammurabi or Mosaic law, the old law has depended, in the end, on the invention of written signs. An old memory that's almost equal to proof, the word 'code' designates at the same time law and symbols or the rules of writing itself. A prior law doesn't regulate an emerging recording medium, rather this law itself emerges from such media. Let's not take the effect for the cause. The medium prevails over the traces it carries. Bereft of sails, steamships couldn't care less about the windrose.

Cognition: The example of memory

Changing historical times, habitat places and laws doesn't, lastly, leave humanity invariant. One more way of interpreting the gesture of storing: depositing information on a parchment, printed paper or an electronic medium consists in constructing a memory, again a concentration. Our ancestors resembled the actors of today who can recite thousands of verses and lines by heart. Such feats now surpass our capacity. As we construct high-performance memories, we lose our own memories, the one philosophers called a faculty. Can we really say: lose? Not quite, for the body deposits this old faculty, little by little, into these changing media; cervical and subjective, this faculty becomes objectivized and collectivized. A stela of stone, a scroll of papyrus, a page of paper, these are material memories suitable for relieving our own, corporal, memory. Already true for libraries, this becomes even more so for the internet, a global memory and collective encyclopaedia for humanity.

A few centuries ago, griots or aoidoses, the interlocutors of a Platonic dialogue, Jesus's apostles, even a student of the Sorbonne in the Middle Ages, could recreate, years afterward, without missing a syllable, the words of the teacher or of a reciter heard in their youth. Safe from the errors of over-intelligent copyists, oral tradition traced a more certain path than written transmission. Our predecessors therefore cultivated their memories and had refined mnemotechnic strategies at their command. As we took notes or read printed material, we didn't so much lose this faculty as we deposited it in books and pages. Just as the wheel set sail from the body, from the ankles and kneecaps in rotation while walking, so storage of information set sail from ancient cognitive functions. Unlike animals, locked into an organism without any secretions of this type, we continuously pour our corporal performances into tools produced starting from such performances. We are losing memory because we are constructing multiple ones.

Losing, gaining?

We rejoin here the ancient and modern tear-shedders, whose discourses and texts lament the loss of orality, memory, conceptualization and so many other things prized by our forebears.

In returning to the snows of yesteryear, let's not hesitate to restage the very process of hominization, such as it was described by the prehistorian Leroi-Gourhan, for example. As our distant ancestors, he said, rose from a quadrupedal position to an upright posture, an evolution that no doubt took thousands of years, their front limbs lost locomotion. Certainly, but the hand gained new performances: for taking presupposes a dedifferentiation thanks to which this organ gradually became the organ of marlinespike seamanship or carpentry, surgery or the harpsichord, the ruler and the compass, prestidigitation ... But as soon as both hands gave themselves over to this refined prehension, which conditions comprehension, the mouth, prognathous up till then because teeth set forward favoured grasping, in turn losing this function, became set back so that the facial angle changed; the cranium became reshaped, freeing the anterior spaces, where the brain could develop frontal lobes ..., and the mouth began to talk.

The balance sheet of these changes lets little losses appear (bearing weight with the front limbs, prehension with the lips and jaw) against profits bearing no relation to these losses (multiple fabrications by the hand, diverse and refined languages, dialogues and objects). As a result, it would be better to say: the arms were freed from the crushing obligation of bearing weight; the mouth was relieved of the tiring necessity of taking; the hand became expert, and the brain reflected. When an impoverishment brings about a better investment, the loss of a function signifies instead that one frees oneself from it and invents the new.

Thus the loss of the memory, during the era following the one when people were singing Homer's poems by heart, freed the cognitive functions from the merciless burden of millions of verses; so geometry, the child of writing, appeared in its abstract simplicity. Likewise, during the Renaissance, an even higher loss relieved scientists of the crushing obligation for documentation, called at that time doxography, and suddenly brought them back to naked observation, which caused the experimental sciences to be born, the children of printing. On balance, the profits win out transcendently over the losses since two sciences, which permitted understanding the world, were born under these circumstances. So knowing no longer consists in remembering, but in objectivizing the memory, in depositing it in objects, in making it slip from the body into man-made things, leaving the head free for a thousand discoveries.

It took me a long time to understand what Montaigne meant when my teachers required me to write an essay on his famous phrase: better 'a well-made than a well-filled head'. Before being able to arrange books in their libraries, Montaigne and his ancestors, the erudite, had to learn the *Iliad* and Plutarch, the *Aeneid* and Tacitus by heart if they wanted to have them at their disposal for contemplation. The author of the *Essays* thenceforth cited them by merely remembering their places on the shelves in order to consult them: what a savings! Consequently, the pedagogy that this Renaissance hoped for would empty the head, formerly full, and would shape its form without concern for its content, thenceforth useless since available in books. Freed from memory, a well-made understanding would turn towards the facts of the world and of society in order to observe them. In reality, Montaigne was praising, in this maxim, the invention of the printing press and was deriving educational lessons from it.

Like bent old men, the children of today no longer even remember the show they saw on television the evening before. What vast science is this other loss of memory going to promote? You can already learn this recent knowledge, or at least visit it, on the internet, such as the new forgetfulness has already shaped it. Yes, the encyclopaedia, whose global network streams with singular pieces of information, has just changed paradigms as a result of the new freeing. Our cognitive apparatus again frees itself from every possible remembrance in order to make room for invention. So we are delivered up, entirely naked, to a fearsome fate: free of every citation, liberated from the crushing obligation for footnotes, we find ourselves reduced to becoming intelligent.

The man without faculties

This long reasoning continues on, invariant, for the other cognitive functions. Calculators and a thousand ad hoc software likewise free us from as many operational functions. The old imaginary slides, in part, into the virtual on the iridescent mosaic of the pixels of our screens. The new technologies render the old cognitive faculties collective and objective, faculties we believed to be personal and subjective. We lose the latter (and how is it that we still have some of them?), we gain the former. The very process of hominization consists in this continuing slide, in this passage, in this loss, in this inexhaustible transubstantiation ..., which is corporal for the wheel and tools, and cognitive for the recording media of the memory, of reason and of the imagination. Let's no longer reason as though the psychology of faculties were still true. This moment of hominescence breaks it forever. What magic mirror, what alpinist's or miner's headlamp did the philosophers who had invented such faculties therefore have when they claimed to explore in detail the dark couloirs and summits of the human understanding? To parody Robert Musil, I gladly name the one who is being born this morning: the man without faculties.

I'll take up again an example already cited: since Galileo at least and on up to and inclusive of my youth, a person devoted to the disciplines of physics, chemistry or biology measured, over the course of an experiment, some variable of some exactly cut-out phenomenon with an eye to drawing from these few pieces of data, rare all in all, general consequences, whose theoretical character ensured application to multiple other cases. A kind of principle of the minimum and the maximum functioned here: few pieces of data for lots of results. Thus issuing from a small number of observations, the law of gravity applies to the Universe. As soon as, become automatic, the apparatuses of examination and measurement multiplied the data in nuclear physics or biochemistry, these data grew tremendously and moved into large numbers, from the giga level to the tera level (10^{12}) and tomorrow probably to 10^{15}. Kilometres of CD-ROMs piled up high wouldn't be enough to contain them. This is a part of the library of the world or a few stammerings of its language. Instead of the experimenter gathering a few hundred measurements, instead therefore of his faculty of remembering, here is an Everest of objective engravings. That was regarding memory, and this now is regarding rational operations. For who will process this vast mass of signs? Answer: only millions of computers connected to each other could do it by working on a common project. The question *who* then requires new answers. Who remembers? Who writes? Who processes and calculates? Entire server 'farms'. Everything slides from the subjective to the objective. I love the fact that computer scientists call these clusters and their connections farms, in the old style: what then does the farmer work on?

To the *cogito* in the first person, two declensions of the same verb 'to think' are added today: the *cogitamus* in the second person of the physicists, which is dealt with by the sociology of science, and the *cogitat* in the third person of computers. The old cognitive faculty becomes decentred in and by a game for three: me, you, that; us, you, them. Had philosophy forgotten that our languages have at least three subjects, that the object itself and the other person each have a pronoun that lets them attain the status of subject? Here are all three at work. Who now thinks the world without the help of others and the aid of computers? That said, the *ego* still has the sovereign power of withdrawing from the game. Who am I? A potential withdrawal. I can *not* play. My consciousness sometimes advises me to withdraw, whatever the cost that must be paid. I am my detachment.

As with the Renaissance, a new science and a different culture are occurring, a science and culture whose grand narratives produce a cognition that reproduces them, enriching them in return. This change of understandings took place several times, for example when the abstract models of geometry or experiments in physics occurred and when information technologies changed. Thus History follows the recording media.

The execution of Saint Denis

When the raging soldiers cut off his head and the latter fell to the ground, he bent over, quite decapitated, to pick it up, then held it for a moment in his raised arms. This fearsome gesture caused, it is said, even his persecutors to draw back. Thus legendary actions, according to Gregory of Tours, recount the martyrdom, around 250, of the Bishop of Paris, named Denis. You can see this scene in the Panthéon, portrayed, in 1888, by Léon Bonnat, the pretentious painter.

Gathering flowers, picking rocks or clods up from the ground to examine them are things we sometimes do, and they presuppose that we perceive them first, then that our body squats down and bends over, lastly that our arms bring them closer to our eyes; they presuppose therefore that, as the seat of the gaze, of hearing, of smell, of taste, of the tongue that speaks and of the brain said to sort and decide, the head acts, in all, as the supreme reference since it seems to command to bend down, to grab, to bring what captivates it closer to it. This deciding authority, judge or leader, is called by philosophy: subject. What lies about on the ground and which our fingers squeeze is then called an object, which, if they can, the hand takes and the subject understands. The execution related here marvellously transforms this figuration, so ordinary, of exercising perception and knowledge since the object to be picked up, brought closer to the tribunal for examination, itself becomes the judge there, exactly the leader or head, and since the fingers that grab it present it to an acephalic deciding authority. What saintliness allows decapitated Denis to locate his head on the ground?

The object, poorly recognized for such by the terrified assembly, suddenly rises above the murderous and transfixed gazes: yes, the very head of the victim held by his hands, lifted above his own headless corpse, still remains a subject. But what other head, absent, sees it without eyes, sniffs it without smell, hears it, without hearing, chattering its teeth and sobbing with pain, which, without a brain, judges it and, without a mouth, proclaims it? Blind, the phantom-head looks at the real head, separated after beheading. Do you see, lastly, the naked and empty subject, without any faculties, that Bonnat painted in a bright nimbus of transparency, facing the objectivized cognitive?

To whom or to what can you compare your console, your computer and its vast memory, its image screen, its powerful calculating speed, its lightning-fast classification of data, to what well-filled and well-made head, maximally dense and brilliantly fabricated? To what transparent light can you compare, then, your own empty head facing these faculties materialized under glass and plastic, in silicon and optical fibre? With everyone having become Saint Denises, now every day we catch hold of, to make use of it, this well-filled and well-made head lying in front of us, bearers of an empty and inventive head on our necks.

The other sense of the verb *perdre*[5]

What a strange power the human body has to transform in parts into objects! We populate the world with tools in the form of a fist: sledges or hammers; of an elbow: levers or pulleys; of an eye: magnifying glasses or telescopes; then of a thousand combinations of functions become, outside, unrecognizable; we even measured them with *pouces, coudées* or *brasses*[6] [inches, cubits or fathoms] … without ever wondering how these machines left our organisms. To my knowledge, there is no explanation that accounts for this leakage or *perte*, taken then in a second sense; luckily, myths or hagiographies make up for this lack on the part of rational theory. For the body *perd* or leaks like an old tapped cask. The way the bishop's body loses [*perd*] its head before picking it back up, the body lets fragments, scattered limbs, leave it, which immediately transubstantiate into technological objects or into substitutes. Before we explain corporal functions and the organism by means of machines, the devices themselves set sail from the body; this ceaseless loop feeds back into itself. The only animals whose bodies leak, humans produce technologies that hominize them in time. Therefore the sudden emergence of the new soft technologies marks a moment of hominescence and changes knowledge here the way it did reproduction above. Through these leakages, which form a world evolving outside our bodies, our physical performance, when it's a question of ordinary energy, and our cognitive performance, when it's a question of information, therefore become transformed. As a result, individuals change, at the same time as the exchanges necessary for their lives, but so do transmissions among the collectivities.

The education of the neighbour

Among the flows of information, one of the most decisive ones for the sequence of time transmits the tradition of one generation to another, which will, of course, disobey the preceding ones, thus happily highlighting the contingency of our fate. Pedagogy therefore always changed at the same time as the recording media for information; this is how, in their times, Greek *Paideia* and Renaissance education arose.

Today, the investment required to set up a university campus, another concentration, with its buildings, libraries, lecture halls, laboratories, restaurants and residence halls, outmatches a hundredfold the costs demanded by the distribution of the same education by means of the new technologies. On account of these modest costs for a multiplied effect, since these networks know no borders other than those of language, these technologies therefore give destitute people and impoverished collectives, who still have little and poor access to the sources of knowledge, their chance. And what, besides, does the word 'university' mean if it doesn't attain the universal?

Who is my neighbour? The woman, child or man at risk of a premature death. Should they suffer, I will judge them in advance to be sisters and brothers, sons and daughters of my holy family. We know that knowledge conditions the lifting of hunger and destitution, sometimes of pain. Given the choice, for one's grandchildren and neighbour in danger, between education and wealth, who wouldn't choose the first one, less unstable than the other sometimes corruptive, more promising, at times creative one?

A project, lastly, in superabundance

If a project worthy of this name exists today, here it is: putting the free sources of knowledge, liberating because more than gratuitous, at the disposition of the most destitute, according to their desire and not ours. What does 'liberating because more than gratuitous' mean? That knowledge does not amount to some good, to money or exchange, but to a strange gift whose mystery we haven't yet penetrated: for if you give me ten euros or some bread, I have them now, and you don't have them any longer; this is a zero-sum game; but if you teach me a theorem or a poem, I receive them, but you keep them; consequently, an addition replaces subtraction; better yet, by reciting the poem or explaining the theorem, you'll be sure to make them grow in you; whence a boundless profusion that no exchange can produce. In the enchanted world of this miraculous superabundance, we practice a multiplicative sharing with the neighbour: a game in which everybody wins.

This is why I have worked for more than ten years to promote distance education, to offer it to everyone, but particularly to the third and fourth worlds. The inevitable pedagogical consequence of the new cognitive order, distance education will distribute more equality into a world democracy that's still non-existent since this name today conceals the most implacable of imperialisms, an energy, informational and financial imperialism.

13 THE END OF NETWORKS: THE UNIVERSAL HOUSE

The philosophy of the address: Enlarging upon the digression

There can be no address without place – this has been the situation from the Neolithic all the way up until this morning, when everything is changing with the laptop and the mobile phone. Belonging to the same family as 'address', the words 'direction' and 'correction', already used, derive from the Latin adjective *rectus*, which, at least here, means straight line and being on the right path: these routes head towards the place mentioned. I therefore recounted one of our oldest stories earlier: once agriculture and livestock farming were invented, when the human group, abandoning the wandering implied by hunting and gathering, lived sedentary in the neighbourhood of the new food source, customs became stay-at-home, and giving someone's address was enough for anyone to find him. To the close relation between a person and his name is added the relation that ties him to his residence. Here humans are fixed to their places.

The word 'address', as I just said, comes from the king (for the *rectus* from just now gives us *rex*), who, reigning here, delimits the places, directions and local borders of his power; his police can thus always find where a subject can be seized. The word then endlessly repeats the thing: the king reads the direction of your address on the correct land registry. He can therefore send, without them straying, his agents or the tax collector to you. The same word reappears in *droit* [law] and its rules. By its roots, 'address' takes on a royal and legal, political and juridical tone. I am highlighting here the first words of a family whose network of relations indicates meanings that the term 'address' all by itself doesn't necessarily reproduce in common language.

The delocalization of neighbourhoods and of cognitive values

But addresses bring even better. The more we give them out, the more people we attract into our private homes, located at the heart of a star where the public roads leading there join. The homebodies themselves thus live at home and outside their homes, there and outside there. And, just as much as the sedentary, nomads themselves mark out, in their wanderings, limited to a finite and known volume, the tents of their clan or of their enemies. Thus a place extends to the set of its neighbourhoods. Letters of a clear code even designated, not too long ago, the quarter of the city where the user lived. Who was your neighbour? The set of those near you whose address you knew.

Moving now from space to cognitive activities, the *Rules for the Direction of the Mind*, by repeating two words of the same family, invented the Cartesian link that unites the expanse and its coordinates. What I was calling the correct direction therefore bounces from peasant locations and from political or legal power to technical address and precision of thought. Habitation, power or jurisdiction, rectitude in geometric method, address lists off the wealth of this place where these roots proliferate. The place, the here, therefore affects our actions and behaviour, exterior and at home, much more than we thought. We love our neighbour [*prochain*]; we think close step by close step [*de proche en proche*].

For the first time, in fact, the new mobile technologies are tearing the address away from this place. Not only is it disappearing, but it is becoming uprooted, in the double human and linguistic sense: the word and we ourselves are losing roots and origins. Paradoxically, the majority of those who use the mobile phone in public or on public transport start their dialogues by telling the person they are calling 'I'm on the subway, at such-and-such an intersection or at such-and-such a spot, on the Chemin des écoliers', as though the old habits were dragging a tail of inertia behind themselves; everything happens as though these people were restoring the place that is lacking, as though the place of their passage were still fixed, even though they were only talking from code to code, with numbers replacing the letters designating, precisely, the quarter.

New world-objects

This new 'situation' – can I thus, without being mistaken, call pure movement a place? – is worth, if I may, our stopping there. Absent from the place, uprooted from the local, we can no longer answer the old questions: 'where are you?; where are you talking from?' We sometimes know where (*quo*) we are going, not always where we are passing through (*qua*), perhaps where we come from (*unde*) and let our position go (*ubi*). These latter locatings used to regulate our thoughts or ensure our old certainties. How many outdated philosophies began by analysing the situation?

Are we leaving the local to join the global? But what does this new habitat signify? Are our niche and our house invading the entire Earth? Must we comprehend that our presence occupies space universally? Are we becoming walkers in the pleasant manner of Rousseau, travellers in the poignant fashion of Pascal: tragically, will we feel ourselves to be wandering, lost, uprooted? Will the king and the law no longer catch up to us? Do we no longer have any rules to direct our minds? These new technologies make us inhabit and therefore think differently.

If I call a constructed tool for which at least one of its dimensions attains the extension of one of the world's dimensions a world-object, the laptop and the mobile telephone, in fact, attain global space and real time: therefore these are world-objects. Connected to a set of points that is equipotent to every place in the world, we communicate at the speed of light. Who is my neighbour, my fellow man? Virtually, the entire human population. In numbers, space, time and speed, these world-objects lead us to live and think outside and beyond place, towards the Universe, which, precisely, has no address.

Once again, the law, the king and rectitude

If the address carries with it this network of deep roots concerning meaning, the king and the law, the correction of method or the uprightness of conduct, is this sudden disappearance of place and spatial direction going to erase all rule? The answer to this question is still undecidable, for, quite the contrary, henceforth it will also become possible to locate our presence wherever we may be. The greatest freedom is often paid for, as here, by the narrowest constraints. For when a criminal inhabited a place, he could flee across all the space outside his address; it has even happened that the police never found the fugitive because he had emigrated abroad, far from the scene of the crime. Only God had the power, through his invisible omnipresence, to find wandering Cain wherever he was hiding, including in a tomb. Yet it suffices today to have the numbered code of his cell phone to track down his mobile trace. Does everyone, in this new space, become God, that is to say, omnipresent? So have we abandoned king and law only to have them conversely return, more powerful than before, quasi-divine? We therefore don't yet know what type of law the activities we are engaging in on the internet are heading towards, an internet become, as I said, a remarkable lawless space. Great changes often begin in these lawless spaces. Whether a new law, practicable or not, arises or not, we have to rethink space, place and site, habitat and relations, the collected objects, the collective subjects ..., that is, the entirety of thought, knowledge in particular, behaviour and ethics. As soon as one leaves the here and now, everything begins differently.

The address towards very old places

How? Reduced to reciting improbable legends, we may never know how a woman, imaginary and initial, invented agriculture; but ever since we discovered a few traces concerning these behaviours, we have known that place, site, niche or house were founded on the tombs of prior ancestors. Where, according to Victor Hugo, does Cain, a fugitive, take refuge, without home or hearth, when precisely he loses all place? In the initial and fundamental spot itself, where place is recreated: the tomb. 'The eye was in the tomb and stared at Cain.' Jean de la Fontaine's 'The Farmer and His Sons' recounts the literally founding scene even better; the paternal old man who teaches to work the land is identical to the one who lies beneath it. In saying these words, in his house itself, he is getting ready to join with its foundation. Farming, burying, founding the place and inhabiting unfurl a single and same ancestral gesture. Close relations, the place and the body all join with death: the loop of this book.

The Latin *pagus*, a ploughed square, says these acts and these things: an old word whose Indo-European origin also designates the stake driven in to delimit its boundary, the funerary stela from which it originated, the tombstone and the dwelling built around the penates and lares (pagan gods or peasant ancestors to whom the house and field were consecrated), the serene peace surrounding the landscape like a nimbus, lastly the page on which the day before yesterday I wrote this venerable prehistory, the very prehistory that, along with the *pagus*, we have just left in favour of the screen. The address goes towards this heavily overloaded place, well before the law and the king arise, towards this *dasein* enrooted in the *here lies*.

In contemplating, now, this agriculture (begun in these places by the body and these funerary behaviours) and this way of inhabiting in common with others, we see that we are living through the end of the place or of the local, the end of the *hic*, of the here and of the now. We no longer inhabit in the sense we have given to this word and this behaviour ever since that Neolithic woman, brilliant and imaginary, invented farming in the fields above the tombs of her forebears. We no longer inhabit in the same way; the majority of us no longer farm at all; we no longer bury our fathers with the same rites; we no longer know how things are with place. We no longer stay in one place, attached, in the past and recently, by the glebe and death; we have just become unglued from it. Observe the land around you all the way to the horizon to learn to what extent our administrators, forgetful of agriculture and in a rush to catch the next plane, have let the land slip into ugliness and be invaded by a chaotic noise. The *paysage* [landscape] has forever left the ancient *pagus* and the set of meanings this word enveloped.

The immortals don't have an address

The end of agriculture and addresses, the end of places and houses, the end of sites and landscapes, must we therefore, against all appearance, announce the end of death? What are we to do with the remains of the dead if we no longer farm the glebe, their archaic residence? Yes, the new global human, the one who now communicates with the four corners of the horizon, beyond the seas and the emerged lands, whose neighbour answers from the other hemisphere, the one whose behaviour affects the hole in the ozone layer thousands of kilometres high and whose nuclear work has consequences that will endure for millions of years, this very human no longer has any but one project before his light ashes melt into the light air: immortality.

We have quit land, place and residence; we have lost the stela beneath which our parents sleep. Which of us has followed the flowing of the slow furrows? Who still knows how to vary the *pagus*, the landscape and its archaic beauty? Said to be domestic, our animals have left their odorous stable; they too no longer inhabit. With the same movement, we have forgotten the *here lies* and the address. Goodbye to place, goodbye to death. A philosopher of the previous generation, an anthropologist of a vanished civilization, was right to link *dasein* and being-towards-death. The reaper haunts place, holds it, keeps watch on it, delimits it, defines and encircles space and time. You will not surpass this bound; you will not pass this period of time.

When we have carried our finitude on our shoulders ever since we drew the foundations of our houses and our fields, how can we believe we have only discovered it today? We have just, quite the contrary, surpassed for the first time the ancient boundaries of place and become unglued from the *pagus*. Formerly tied to a stake like goats, now our acts and works, our scope and range reach the Universe since we produce world-objects reverberating in global space, since for the first time the neighbour [*prochain*] becomes unglued from the near [*proche*]. That was regarding the expanse; what then on the subject of time? We likewise cross its borders, formerly reputed to be uncrossable. Life expectancy is gradually lengthening according to numbers our predecessors could never have imagined. Will we know how to decipher the apoptosis signal? Will we know strange autumns without the fall of dead leaves? Among the ancestral division of things that depend on us and things that do not, death held the major post. Unshakable up until this morning, is death going to change camps? Not only have we partly won responsibility for our health, for our deadly risks consequently, but on the day these signals form a language that's finally comprehensible, will we ourselves be forced to choose the solemn hour?

Many cultures define humans by humus: autochthonous, come from the earth, nourished by it and returning to it to nourish it. We have just left the humus, the one we tore up in order to feed ourselves, the one on which our residence was supported, the one that welcomed our body at the end of its adventure. Does this mean we are leaving the human? Yes and no, for all these events mark a decisive moment in hominization. We will no longer be humans the way we were; we are becoming humans the way our children will understand it. In the meantime, we are autonomous, our own king, our own law; free, perhaps, in the time and global space in which we wander, to give death our address some day.

This new relationship to death ensues, from afar, from the victory over death Saint Mark (16:6–7) celebrated at the dawn of our era. Bearing vases of spices, Mary Magdalene, Mary the mother of James, and Salome, holy women, arrived, while the sun was rising, before the open stone and empty tomb, by the entrance of which stood a man dressed in a white robe. 'He is risen', he tells them. '*Non est* hic, ecce locus ubi *posuerunt eum.*' 'He isn't *here; behold the spot* where they laid him down.' Empty site, no more place. Death was already absent from space and time. You will find the son of man everywhere and always elsewhere. Here, a dismissal of charges due to no grounds for prosecution[1] is issued for all humans: without address, exculpated, we are saved. From death?

Landscape and the cost of places

The relatively sudden passage from place to non-place, from the local to the global, from the square of alfalfa to the global horizon, from the tomb, where what has no name in any language is forgotten, to the ashes scattered to the four winds …, from the near to the neighbour [*du proche au prochain*] … causes what our parents called the landscape to dissolve around us, a landscape constructed and understood as a propagation of the *pagus*: next to the cultivated field, other ways of farming unfurl in proportion to the strength of the people and animals working in the vicinity. This stitched-together tatter, with elements of various forms and polychrome hues, running towards inhospitable lands, has long presented the precious marquetry of Eurasia, Africa and Latin America, agricultural countries by choice, not including mountains, deserts and swamps. I have in the past, from the very word *pagus*, tried to extract women and men, works, tools, laws, rites and peace, in short, the set of the anthropological behaviours from which the peaceful beauty of the landscape [*paysage*] issues, in a mournful song that only someone who has lived and sweat in these swallowed-up places could still compose. Painting bulls and rivers, Potter and Ruisdael, genuine peasants, thus made their land attain the peaceful ecstatic while the history of the seventeenth century was waging war around them and spreading carnage, ugliness and chaos.

Built slowly since the Neolithic, this landscape vanished first because connecting engineering works, tied to the elimination of the peasant population through cost prices and the imperatives of profitability, tore up the copses and hedges, filled up the ditches, the ancient boundaries of the medium-sized *pagus*. Covering over valleys and hills, burying forever the ancient landscape underneath their monotony, vast gauche surfaces reproduce corn and sorghum in dismaying colours as far as the horizon. Required by tractors, the growth of these planar dimensions decreased rural density while increasing the sown land: a landscape swallowed up at the same time as the peasants, now devoted to solitude and isolation.

This new landscape, secondly, boils down to urban ways. For the landscape, which the inhabitants of the cities always talk about without really knowing what they are saying, remains an agrarian framework for thought, for human proximity, for behaviour and aesthetics, presupposing a peasant culture; when this peasant culture predominated, there in fact existed, amid the sea landscaped in this way, rare islands called villages or cities, where the political dominators lived; even though they held power, they still had need, in order to eat, of this landscape into which their habitat was plunged, dispersed. But the hominescence whose sudden blow I am describing in these pages has completely reversed this state of affairs: the city is devouring agricultural space and invading it instead of propagating itself there, forgetful of the very meaning of this latter verb. Literally, the city is politicizing it; yes, like Potter or Ruisdael, the peasants not only lived outside politics, which, all things considered, reduces to the tautology that they weren't living in the city, but also and especially outside History. Inventors of this history and good at politics, the inhabitants of cities, at least Western ones, for I can only talk about my experience, still had to eat. Yet today they are dictating dietary rules: they are imposing a food that's abundant, not expensive, aseptic, without provenance, risk or flavour.

Pages of peace

And they are dictating them so precisely that they cover space over with their writing. The peaceful landscape of yesteryear and yesterday, woven of *pagi*, maintained the silence twice over: without any sound or voice, it was sometimes pierced, at rare moments, by neighing and barking, by children's cries or the chimes of the Angelus; it was particularly unaware of writing. We only heard our neighbour: a few bursts of conversation, children's shouts, the lowing of cattle. Around the illiterate people a silent space reigned: the written text hadn't yet polluted the world; it only lay in town, on the shop fronts and the corporation signs. Now, its explosion is befouling the volume as far as the horizon; from ankle-level all the way to the mountaintops, the eye can no longer perceive the form of the topography nor the hues of the dawn since the irresistible attraction of reading incessantly subjugates it to empty messages and advertising bombast. At a time when we are no longer reading books, the entire world is taking to the page. At a time when peasants are no longer meticulously sculpting the landscape, dominant writing forbids seeing the evil it is doing to space by diverting all eyes from it: sucked up by the letters of a continual text, we don't perceive forms or colours. I didn't suspect just how far the literal ugliness could reach before seeing France, my oh so beautiful country, submit to imitating American streets by lighting gigantic words amid hideous colours. Departing from the cities, this propagation without *pagus* but with page is destroying their entrances, now given over to the worst degradations, so as to now invade what remains of the countryside. Better, forms and topography are losing the third dimension in favour of flat surfaces, the recording media for these letters. A witness to the lack of culture of the new dominant classes, noise trash, in addition, is accompanying these visual garbage cans. Passing by train along what one could still, a century ago, rightly call the Côte d'Azur, Mérimée recounts that women fainted in the train compartments due to the power of the fragrances emanating from the landscape.

Just as electricity made the eye forget the delicate perception of gradated twilights, so engines exclude the ear and nose from their refined functions. Exhaust farts and toxic gases eradicate hushed messages and plant fragrances. City economies paralyse the five senses of *Homo sapiens*, a term whose forgotten translation specifies that it had to do with a person of good taste. With sensations stifled, this person is dying, giving way to *Homo clarinans*, connected to the musics of the world like a cow to its bell, in order to avoid the ignoble noises of engines. Should a storm now blow over this dead landscape, no one can pay for cleaning up the forests entangled with trunks and branches; while the poverty of the peasants allows city people to pay less for food, the city folk, living off this slavery, don't want to pay for work on the landscape at the cost of city salaries, thereby admitting, once more, that there can be no landscape without peasants [*paysage sans paysan*]. As a result, a chaotic disorder becomes established, in a land [*pays*] that no longer has any right to this name, on the very sites where an ancient and meticulous work had established order, calm and beauty.

These are the costly prices of the recent passage from the local to the global: the major injuries of the old propagated places. The losers always pay for the victory of the powerful. No one will dress these wounds because the decision-makers no longer dispose of the five senses sufficient for understanding, nay, for even perceiving these local pains. They only see the global. Admire them, on your screens, flying over at times, on a plane or helicopter, the theatres of these catastrophes so that their cameras will let you contemplate, in your turn, what they saw when they went there; diploma-holding administrators, the subordinates accompanying them will write reports destined for a drawer.

No one lives on site. No one lives in a site anymore. The power of a person is measured by the number and length of his travels. He occupies space at the speed and in the throbbing of engines. The glory of a person is measured by the noise he makes: this sentence isn't to be taken in its figurative sense any longer but in its literal sense: so his power is measured by the amount of filth he trails behind himself, every gaseous, noise and textual pollution included. Since certain animals define their niches with their pee, power holds global space by means of this filth. To this urban invasion corresponds the end of landscape propagation, death throes which will continue until the hominescence in question finally transforms general opinion and then the leaders.

Revisiting the natural contract

How? This will only take place through an unprecedented and long political and social education, which one day I called the Third Education. Maybe it will be necessary to stop calling works, thoughts, behaviours and actions that can no longer and must no longer be limited to the city or that don't take into account the explosion of the city into global space, political. When 90 per cent of men and women die of hunger, when water becomes scarce and the air unbreathable, when the products of our industries disrupt global temperatures, political decision must be added among the expertises surpassing the single grip of the cocooning hive in which groups live among themselves. On the side of the sciences, sociology no longer suffices for this without geography, nor does economics without climatology; in short, social knowledge must mix with the hard sciences in order to understand and manage a dangerous game that doesn't so much bring two adversarial teams together as it brings these partners together plus their conditions for existence and opposition: every war takes place on Earth. The educated-third thus balances his knowledge, his acts and his behaviours. In French philosophy, such a supplementation has ancient letters since in the past it set Rousseau's disciples against Montesquieu's, whose works took into account latitudes and climates, in a word, the world, place again and space today, in which humans do not live alone but in the midst of animals and plants, without taking into account bacteria ... and amid the inert masses of earth, air, fire and water. In philosophy, in pedagogy, but especially in politics, acosmism is becoming dangerous today for humans themselves. By continuing to live thinking themselves to be alone in the world, the world itself might take all life away from them. Their actions as subjects transform them into passive objects suddenly submitting to the forces of a world that, according to the loop already described, becomes the subject of this action in return. Hence the strange Natural Contract I proposed, little understood even though legible everywhere. But how are we to read or sign it?

Where and how are we to read this contract?

A stamp that draws, that scarifies, that imprints peasant practices on the earth, the landscape draws and marks the contract the farmer has never ceased to sign, ever since the first domestications, with and on the localities of the land. It writes it in lines: hedges, ditches, furrows and rows of vines …; in dots: isolated trees, bogs, wells and boundary stakes …; in pages: the *pagus* itself, so well named. In saying this, it's not a matter of comparisons or images you might scorn under the name of poetics but rather of concrete characteristics since every wanderer, traveller or walker, whatever his culture and language may be, knows how to distinguish a wild moor from a cultivated field and so recognizes on site, deciphers and reads this languageless contract, even if illiterate, even if he has never studied on the page (the child of the *pagus*) lines (children of the furrow) written by pen or stylus (children of the plough). For everyone sees and reads in the ploughed field the first writing, one imitated from the natural writing without any human intervention dug by winds, torrents, volcanoes and glaciers on the ground; everyone reads and sees on the landscape the drawing or designation, nay, the sign and signature of this first contract, drawn up in a hundred Asian languages on the rice paddy fields of China or Malaysia, in Indo-European on the sown fields of the Po or Garonne plains, or in some Amerindian language on the high plateaus of Bolivia or Peru. As these landscapes disappear, multiple in their deteriorations although single in this sense, the contract written in this way and continuously re-signed by the vanished peasantry is no longer renewed, and we are replacing it with a new drawing, that of the global or natural contract, sculpted, for the moment and in its turn, but in the negative, by the holes in the ozone layer, the fragmentation of Antarctic ice sheets, the receding of Alpine glaciers or the slow rise of sea level. We thus know a rough draft that will one day have to be, under pain of death, rewritten as a final draft.

Thus and at least this Natural Contract has the advantage over its ancestor, the Social Contract, itself spoken so highly of although virtual as well, of being able to be read and deciphered in the space of the world by anyone, even if uneducated or illiterate; so visible, so evident that one must have had one's head buried in books to no longer see it; as transcendental as its predecessor and yet concrete since, displayed in physical space and in the process of humanizing it, it lies at the origin of writing itself. Can we dream better than a text that's legible before every written language, although this text is only second, after the diverse traces and marks of the elements on the elements: riverbeds on the land, channels, scarifications of glaciers on the rocks, crevasses, accumulations of lava, of erosion sands and even of fossils, memories … Our writing thus only occurs as third generation since the things themselves do it as well as us and on each other, to the point that we have to use things to write on things.

Hominescence renews the face of the earth

So read and see, in the agrarian and primitive text, in the place where its drawing remains, the old draft of this ancient contract, visible and legible on the earth worked by manual tools, as a family and with the help of neighbours, written pages, drawn landscape, local propagations on the surface soil. Decipher, now, at least what I have just called the rough draft of the new contract written up by means of our world-objects, by entire populations, on the planet in volume, and at most the draft drawn by optical fibre or the new networks. From one text to the other, the human relationship to the world is transformed by a different contract signed between two partners, us and an environment that changes scale. Each of these stages thus writes the new step in the process of hominization positively on the Earth.

So I am in no way inventing the moment of hominescence since each of you could, can and will be able to read it on the very face of our planet: 'And you will renew the face of the earth' (Psalms 104:30).

Knowledge, treatments and laws

In the old space of concentrations, the pupil or student, the sick or injured person, the citizen or subject, three common examples, had to travel to the school, hospital or offices in order to attain knowledge, treatments and laws, all grouped together in suitable centres. These poles and roads departing from cities or ports and returning to them formed one or more networks, whose various drawings in fact covered over the wild face of the land and humanized it: a tatter of the *pagus*, a multiple latticework of byways, routes by land, sea, air, post, telegraph …

But, still recently, the three users in question had to, always and at least, travel to their computers, which, decentralized but still fixed, had in their turn taken on the function of centre or repeaters of other poles. In total, it was only a matter of another network, of course denser, more reliable, better organized, faster, but of the same form. Nothing had changed ever since humans occupied the world. So even the internet hadn't yet adapted easily to education, medicine or public services. As long as stations remained, learners, patients and subjects incessantly ran along the roads to the fixed centres where they received knowledge, cures, services in general: at least to their consoles. Up until yesterday, consequently, space was arranged only from the point of view of sources, not in the interest of the users, always tied to places, the ones occupied by their computers; these places still valiantly resisted the new technologies, not without good reasons, since screens and hard drives, always heavy, budged as little as their walls and since roads, still ill-adapted, didn't always clear obstacles or interceptions. In other words, these tools still formed a network point by point and by difficult and discrete roads. The English word 'web' even designates a globally centred spider's web on which a fearsome predator keeps watch for any living being that may pass by. An admission?

Because they are becoming usable everywhere in space, 3G telephones, for example, in giving mobility upon demand, promise to construct a dense and continuous expanse. This is a decisive difference: so, at least in principle, places where an individual can attain everything no longer exist; here or there, a student can travel, a patient fall from an attack or suffer an accident, a citizen want to vote; they all three attain, wherever they may be, the demanded knowledge, the necessary treatments, the polling station, the administration and the laws. Reciprocally, if the flows of information serving intelligence, the body and the collective claim to truly travel to those they educate, treat or federate, they too must be able to run anywhere. Tomorrow we will have to create new roads and improve them again, so much do we demand performance from our technologies that they can't always achieve. Mobility changes space.

The body

Lastly, when, scarcely the day before yesterday, the elementary school teacher put the pupils in front of screens, these young people inevitably turned their backs on the instructor; they lost the corporal contact that's so beneficial to pedagogy. As soon as everyone holds a mobile receiver in their hands, they turn around, and the learners face their educator who, once again, looks at them. Dialogue is re-established. Face, back, hand, the entire body changes attitude. Jean-Louis Gassée subtly said that reading a book gives the body a passive attitude comparable to the attitude of a passenger in a car; with Montaigne or Flaubert driving, I have chosen to have myself taken on board by them. In front of the computer, the body, slightly bent forward, quite the opposite, takes on the attentive and reactive attitude of the driver himself; I take the place of the author. Whoever condemns the new technologies, accusing them of passivity, has never used them. His spinal column would have forbidden such nonsense. The fact remains that the inclination of the back encloses a private locality, quasi-narcissistic, defined by the face of the body and that of the screen. However connected to the entire world we may claim to be, we nonetheless close over ourselves, indifferent to our neighbours. Thus the defenders of the most unfortunate people in the world are sometimes unaware of those close to them. We no longer see our neighbour.

I will take up the example of the 3G phone again. The body frees itself: standing, seated, even walking, I can handle this new tool that only requires the palm and the ten fingers. Ergonomically economical, at least for the hand if not for reading, it no longer mobilizes the entire body. I no longer shut myself up in the closed dungeon between the screen and the blind back; even distracted by my manoeuvres I keep an eye on those around me. Up to here, nothing new with regard to the mobile phone; but, in addition, by means of the internet, I attain, at least in principle, the entire encyclopaedia of the world: information, games, musics, treatments, laws, administrations, sciences, images and landscapes. Even those who speak the French language fluently have long forgotten what *maintenant* [now] means: holding in one's hand [*tenir en main*]. The present, the immediate, that vanishing fraction of time, that rare moment all wisdoms have praised ever since these wisdoms had got themselves talked about, my language says that I hold it in my hand. Maybe I have never held anything in my hand except the present, except my neighbour, except the female neighbour that's present. So why invent the words 'in real time' since we dispose of a better expression, simpler and more concrete? By means of the Universal Mobile Telecommunications System (UMTS), the global projects itself in its entirety, for everyone, into every possible locality at every instant. Each of us, now, holding the world in our hands ...

Dedifferentiated, the hand slaps, caresses, shells or takes; at the end of a tool handle, it multiplies its grips and its enterprises: strikes, chops, carves, drives ..., agricultural, industrial, artisanal, surgical, pianist ..., writer or semaphoric as well since it calls and makes signs, as we see with the deaf-mute. A universal organ for energy, from the shovel to the steering wheel of a car, the hand therefore makes the transition to informational tools, from the pen to the console. So admire it as one of the masterpieces of evolution. By making the sum of the media available, the UMTS attains, at least in our projects, the status of a universal tool of information, while also making the transition with energy since this tool can activate any motor device remotely. So describe this apparatus as a masterpiece of human skilfulness or of exo-Darwinian evolution. An act of an inarguable newness, taking the UMTS in hand therefore consists in connecting a universal tool to a universal organ. Life and technology together tend towards totipotency.

Once again, the relationship to the world: The end of networks

On balance, when the body's gestures change, when we quit the discrete to attain the continuous, fixity for mobility, blindness for the face-to-face, the body itself changes spaces. New students, new patients, new citizens, we no longer inhabit the same niche. For here it's a question of a new occupation of the world. Only yesterday, this person and that person enjoyed having access, from at home, to the totality of their fellows, at least virtually. They didn't realize the vast distance that remained to be covered to get to the universalization they were extolling. Mobility again changes this situation. The at-home is distributed everywhere. Computers or fixed-line telephones hadn't yet detached addresses from places, nor space in its entirety from its concentrations. Now, the sources and terminals expand into a continuous space. Not that we can live everywhere, but that in every place we move through, we dispose of the same services as at home.

The mobility of technologies renders the four questions of place I posed at the beginning – and which worried human fate so – incomprehensible because this mobility dissolves the very form of the network, of the spider's web, of the Web, of the fishing net, of the fabric itself or of the text, with their discrete sites and roads. What remains is a plane with continuously dissolved points. Nothing can distinguish them anymore, wherever I may be, wherever I may come from, wherever I may go and whatever the place I am moving through may be. At the very moment when no one is talking about anything but networks, they are dissolving into a general short circuit. We now live in a space without any measurable distance. Formerly an abstract science, topology describes the space of our habitat quite concretely. Now our house is built less in a geometrical space, presupposing a measurement of the earth, endowed with assessable distances and making the four questions possible, than in a topological space without any measurable distance from any point to any other point.

Written and oral: The stakes of changing recording media

Messages circulate on this new space. In what form? Our fathers wrote to their correspondents; we speak to them. For the two major revolutions previously having to do with recording media allowing the storing, transmitting and receiving of information concerned writing. Leaving the voice behind, these revolutions didn't associate it with their innovations, as though the ever-new written stage prejudged the oral stage to be obsolete or primitive. Yet the contemporary revolution, on the contrary, associates, in a new and unexpected synthesis, the written and speech precisely owing to the change of space.

Known for money, the law of increasing lightness or mobility becomes generalized to all information. From marble to the flying page, writing's new recording media acquire the dimensions of oral transmission, light and mobile. These media even surpass it on its own terrain. The circulation of information begins at the same time as the circulation of its recording media. But, in addition, practices of storing, inaccessible to the voice, are instituted. The oral only knows the body and the subjective memory, a 'faculty' therefore tied to this stage; speech moves with the mouth, its source, but can only be stored in and through mortal flesh, while writing has two advantages: it is deposited, concentrated and moves, including after the disappearance of the author or the interpreter. A new setting sail, different networks by centres and roads, the book kills the oral.

Today, without losing any of the museological benefits of the written, we are once again living in an oral civilization. As long as distance founds human experience, the written prevails over the oral. As soon as the remote, erased, comes nearer to point of touching me in a space without measurement, the voice gets the upper hand again since anyone, wherever he or she may reside, can hear and talk to me wherever I may be going. In a world without separation, what need do we have of roads, horses or power lines for angels, messages and messengers? What need do we have of networks? The omnipresence allowed by the new topological space brings orality back. The omnipresence of music and of certain vocal kindnesses whispered beyond the oceans lets us hear this evident fact.

However, the arrival of the computer had given the written a comfortable lead since this arrival was accompanied by home printers. But writing has to wait: it must be learned – there are more illiterate people than mute ones – before it can be engraved, printed, reproduced, stored, sent, received, deciphered or read. It pays for its mobile faithfulness across space and time with a given length of preparation. It isn't just suddenly invented and delays transmission. For, reliable, writing remains, whereas speech flies [*vole*]: but speech wins when speed is at issue; and our civilization demands this volatility. We live less in the Age of En*light*enment, bathed in its brilliance, than in the age of its speed, informed via its lightning-fast propagation.

Consequently, recently freed from its shackles, weight and wire, the mobile phone gives the advantage back to the voice. The transmission continues, of course, to reach any point in the world but in addition rediscovers the mobility of the source, as during the times of the aoidoses and troubadours. Consequently, it attains real time, which the written can't reach. Its principal benefit, the now of the oral does away with the waiting times of the written. The voice has just conquered again. If, lastly, the written costs more than the oral and language more than speech; dialogue, democracy and freedom have more chances with the latter than the former, and equality returns. What writer has dared to write that inequality among men comes from the fact that certain people write and read and not others, vastly in the majority? Like freedom, equality is decided by the body: everyone talks, much fewer write and read.

Better, when writing emerged, its role was to set down – to steal? – speech. The *Iliad*, Plato's *Dialogues*, the *Gospels* and the *Metaphysics* transcribed pre-Homeric songs, discussions around Socrates, Jesus's parables and Aristotle's courses. Flaubert himself transformed his bellowing declaiming into style,[2] and the journalist dictates the day before, the article to appear the next day. At that time, the voice became the servant of the written and remained so up until this morning, when reading itself always found its vocal or mute source in the printed page. The technology revolution reverses the service: the written now becomes the equal and sometimes the servant of speech. The written is displayed above the microphone, on a small screen, our new page. Who would have said that we would one day hold in our hands this descendant of the *pagus*, which formerly lay beneath our feet, that we would put it to our mouths and ears? Not that we are returning backwards towards what we scorned in oral-traditioned civilization, but we are reversing the synthesis, the very injustice, at least the inequality, on which the culture of yesteryear and yesterday was founded, in which the written prevailed by far over the vocal, therefore certain people over others. The temporary victory changes camps. What person enamoured of democracy wouldn't rejoice over this?

The stages of hominization

To describe the various occupations of world space by *Homo sapiens* in (too) broad strokes, it would be necessary to distinguish multiple stages that are highly differentiated: its emergence and its settling in the East African savanna, its crossing of the Red Sea before bifurcating into Eurasia, its traversing of the Timor Strait before transforming into Australian Aborigines, its traversing of the Bering Sea before populating America, the recent demographic explosion, etc. This gradual invasion founded, here and there, more or less fixed settlements, well or poorly linked to each other, various centres and roads of all kinds. Routes cut off so often that we forgot each other just as often. A hundred not very reliable networks formed, which went about multiplying all the way up to the enormous number of connections we have today. So the occupation of the world by humans has been drawn by the very form of the lattice ever since their beginnings; this form always evolves via central poles and mutual distances. In other words, the invasion of the world by humans has always taken on a form of this type, invariant across its variations, this discrete drawing in which gaps separate places. Obviously, this populating never attained the density of a spatial continuum nor the total nullification of all measure and all distance; yet we are moving, at least virtually, from the discrete to this continuous since everywhere now a pole of possible communication exists. There no longer exists any place in the world without an infinity of roads leaving it and arriving at it and without the length of these routes being nullified. When messages don't pass through a place, we will dispatch them differently.

How is this state to be characterized except as a moment of hominescence? I repeat, a newness occurs as soon as an era that can be defined comes to an end, a newness the more decisive the longer the era has lasted. Here our oldest relationship to space and time, to the here and now, to being-there in some manner is becoming transformed. The word 'density' is also taking on the sense topology gives to a dense space. The landscape had written a Natural Contract by opening up a thousand networks of roads, a thousand kinds of writing. We are now drawing a virtual contract, a white map, on the face of the earth.

On distance

Consequently, the fundamental human experience remained, from our emergence, that of distance, which made us into that *Homo viator* who came out of Africa and spread into the Universe. But why did we travel like this in such an irresistible fashion and so far? In what direction did we begin this immense journey, which today is passing into the virtual? What irresistible attraction was exerted like this on us from yonder? Where was desire spurring us on to? In what direction? Excluded from the uterus at birth, from the breast at weaning, from the mother house at adolescence, from the native soil by profession, from happiness by spurned love, in short, from the fundamental place, often ill from homesickness, we tried to reach the sources of food, of friends, of a woman, or, conversely, we put kilometres between us and the powerful. Vital experience remains an experience of separation: from the familial niche, from the distant princess, from the paradise lost, the unhappiness of absence, the impossible happiness of reunion. The highest wisdom in the world consists in loving one's neighbour, in drawing one's neighbourhood, therefore in measuring the smallest distance possible. Well-named, every pedagogy in the world leads the child to the end of a journey or of an initiation. Distance founds anthropology, our long prehistory, the adventure of journeys, our contingent nature, wandering at random. Without the entirety of these travels, there can be no human experience. Philosophy measures the distance to wisdom.

Our languages provide proof of this since the very word 'experience', derived from an Indo-European *per*, marks out a route in space towards the places we are heading; we find ourselves separated from them. We hold one part of our destiny, and we cruelly lack the other. In disequilibrium, our existence consists in the quest for this part, lying at a distance, impelling us to wander in search of it, desirous, hungering, thirsting to find it. Existential and destinal questions ensue from questions concerning this distance to a place, as though this space, literally metric, conditioned our civilizations, cultures, metaphysics and religions, as though there existed a transcendental geometry, a fundamental network suitable for assessing measurements on our earth, beneath the deepest questions humanity asked. Homesickness always spurs us onward. Thus the West's singular time, begun with the epic of Gilgamesh, leaving on a journey to seek immortality, continued by Ulysses' wanderings in the Mediterranean, contributes to general anthropology since we now know that the entire human race bifurcated at the time of global travels; these concrete journeys, repeated in the teaching traditions, are also reflected in the more or less pathetic metaphysics in which humans search for meaning. Can this meaning be thought in a space without distance? Who hasn't ever suffered from this experience? Who hasn't written its sorrowful translations in love poems? Certain people even added its sum up in order to measure an infinite distance, transcendent, between them and the divine subject of their dilection.

Envoi

Yes, our most recent technologies have forever destroyed this fundamental geometric web, which produces networks and endows meaning. The fact that we now inhabit a distanceless topological space changes our destiny and our philosophies but our anthropology first: we are no longer the same humans. We no longer live together in the same way. As a result, social functions change in detail. But above all, since the close–distant attained in this new dense space can attain the height of alterity, is the global law – love one another – being substituted for the local law, cruel because it produces exclusion and conflicts (love one another only inside one's group), a law closed over the family, the reproductive line, the country and its landscape, the region and its address, in short, spatial proximity and its absence of distance?

If the word 'existence' does mean this deviation from equilibrium that one day set us on the road and in motion for so long, we are no longer living the same existence. Barring one distinction: the absence of distance concerns messages, not yet all the functions of the body. We attain everything except the beloved body. All of anthropology changes, but this transformation leaves acts of desire intact, acts that are just as singular as they are integrated in their mystical sum. Everything of course varies, but the gap of love remains. *Honor/honneur, horror/horreur* …, the French language translates Latin words ending in – *or* with terms ending in – *eur*, with one notable exception: *amor* and *amour* [love], whose ending marks its troubadour origin. These musician-poets spoke Occitan, my dialect of birth, few words of which end in – *eur*. If a shred of pride in local belongingness remains in me, I owe it to these ancestors whose genius invented this word and this thing, to which the majority of my philosophy refers. This love, as virtual, always moves towards a distant princess, my neighbour.

14 THE THIRD LOOP OF HOMINESCENCE

Denunciation and announcement

Ethics again. Whoever denounced their neighbour, during my youth, was taken to be a bastard; we no longer frequented an individual who was as low as he was dangerous. A few decades later, no one could be taken to be endowed with talent if he or she hadn't denounced, with much noise, some public or private scandal. Immoral yesterday, now moral, this so-called transparency nevertheless issues from the same act that unveils a hidden crime, whether true or untrue. The only change concerns the scale: in the one, a couple of kids, in the other, millions of citizens.

Far from preceding a possible trial, denunciation replaces it and punishes swiftly. One French prime minister committed suicide after a media campaign of malicious gossip; humiliated in this way, how many contemporaries would have acted as he did? Many, myself perhaps, the accusers as well. Denouncing in a newspaper, on the radio, the television or the internet invites too great a number not to turn into a lynching. A regressive and archaic behaviour and which only seems just for large numbers. The accused is taken first to be guilty and even in his own eyes: his guilt doesn't ensue from any objective establishing of a crime or an offense but from the opinion, quickly unanimous, of the crowd that burns him.

Large populations

For truth is defined in two ways: either by the thing itself or by consensus. Yet, as science and philosophy do when they persist in discovering the true, justice seeks, in order to find a guilty party, to avoid a heated, quick and shouting persuasion of the masses in order to organize a cool and slow consideration, by a selected jury, of the case in itself. So truth no longer depends on the number of people it persuades, nor on a consensus (at most dangerous, at least conventional), but on experiment, evidence, testimony, proof. But, once again, isn't a collective labour essential to supervise these procedures?

Far from accusing the current practices of denunciation by name – heaven forbid I should denounce – I simply think that the new communication technologies allow us to work directly on this large number and that everything depends on this number objectively; such an innovation doesn't happen without social, political and judicial consequences and demands elucidation.

Society machines: Large numbers

Teachers and other orators know that the content and form of speech depend on the number of listeners; rhetors don't address ten, fifty, two hundred or a thousand people in the same way; by successive stages, the nature of the message broadcast, the quality of the transmission and the ease of reception change. Hold, by one end, a kind of stick: short, it permits writing and drawing with exactitude or carving clay or wood finely; longer, its action becomes less precise; beyond, it takes on an autonomous swaying whose vibrations escape your control; on this side of the handle, you therefore drive it more or less easily depending on its length; and this is how it is with the group and the crowd, seated or standing, closer or farther from you depending on its volume. For this stick, become here a kind of social tool, is controlled less and less the more it lengthens; and this goes all the way to extinguishing all possibility of acting beyond certain limits. Yet, for the first time, the big media are passing them; for the new channels are opening up direct access to large populations. A political tribune never declaimed in front of millions of citizens; a theatre actor, a trial lawyer or a professor likewise never addressed such a quantity of listeners or spectators. While talent for public, scientific, religious and judicial speaking adapts to these stages by varying according to the number, these changes of scale now exceed possible rhetorical experience and reach different levels we haven't mastered as much.

Up until the middle of the last century, we didn't know, except for lampoon, anything other than direct and live discourse as a social tool, as a machine suitable for 'working on' the collective. And no one could talk about machines regarding language, voice, eloquence or the influence of the word. The orator's body had to be sufficient; he had to 'hold' the hall with his powerful 'set of lungs'. Hence the importance of the speech sites, rhetors before the Areopagus, military harangues, tribunes under the rostra, ecclesiastical or Sorbonne chairs, judicial eloquence, election campaigns ... For, from site to site, the source or the transmission varied across time, but the scale of reception remained stable.

We never realize to what extent phenomena, their nature and their laws transform when their scale changes. We look down on quantity even though its increase rarely leaves quality invariant. Examples: confectionary artisans construct, during festival times, Eiffel Towers made of rock candy that would collapse if they exceeded two metres in height; in order to keep the same form, the material must be changed and moved to steel; likewise a frog that inflates itself to the size of an ox must become an ox if it wants to survive and can't stay a batrachian without exploding; in order to preserve its living material, it must change forms. Thus I can explain the exquisite sophistication of certain disciplines clearly to twenty, to fifty students; I summarize and become more unrefined above a hundred; beyond a thousand, my discourse loses even more of its subtlety; but after five million, who are absent to boot, what rhetor will master something whose laws he doesn't know and for which there isn't any rhetoric? In the first case, I work less on groups than on understandings; the effects of my course on their local number will no doubt leave society invariant. But in the second one, it's already a matter of society as such, no longer of the rare group, but of the global collective. So by reaching large numbers, the new technologies are already shaping the collectivity as such. To speak directly to the social: what does this expression mean, what does this profession consist of?

Voice instruments: The two social loops

It is a question of machines; it was a question of institutions. For, visit the city: the ruins of Athens or of Knidos in the fifth century BC, the Vienna or Helsinki of the nineteenth century, the Paris of the Second Empire, the Tokyo or Rio of today; Senates, 'Parliaments' or 'voting' spaces, theatres and Opera, Stock Exchanges and auction halls for shouted bids, tribunals, chapels, churches and temples, classrooms, schools and campuses, public squares, agoras or marketplaces, cafés, even partitions perforated with wickets ..., these are buildings carefully designed, fashioned and organized for the voice. Certain ones deviate its speech towards politics; others towards commerce and for exchanges of money, still others into law and jurisdiction, culture, education or religion. The architecture of the theatre – stage, tiers and orchestra – allows the collective to see itself discourse and rhetorical eloquence to resound so that everyone can hear and respond; this form varies fairly little from the Chamber of Deputies to the medical lecture hall and from the cathedral to the courthouse, except that the confessional and the window counters of administrations and banks guide the whisperings towards a more tightly channelled dialogue. Does society organize, in these various and parallel buildings, voices into parliaments, votes and shouted bids, or does the voice itself organize society? We can get out of the debate by invoking a reciprocal causal loop.

The effectiveness of the voice didn't come immediately from its meaning but from its sonority; music therefore – intensity, rhythm and melodic envelopment – fascinated the group first and still enchants it, a group glued together by the magic of this living cello. The eloquent periodic sentence induces the listeners to slowly climb the slopes of a sea swell propagating across the crowd and whose tremendous bulging hoists it towards a peak from which, ecstatic, it contemplates an inaccessible horizon and whose descent into the trough of silence and hypnotized attention causes to fall into another world the old group torn apart by debates and suddenly united by new bonds. Concord in the sense of acoustical harmony mysteriously brings about a social contract, at least temporarily. So the propagated waves seem to be independent of the threnodies of eulogies, of the hopes for peace begotten by a programme of foreign policy, of the innocence of some pitiable defendant, or of the depth of speculative analyses since the architectural form remains almost invariant for rites, voting, sentencing and lessons. These voice machines – court, theatre – first and foremost collect heat and number, interference and the flight of sound waves. Said institutions are therefore built like acoustic instruments, tools themselves local, for the weakness of the human voice imposes the well-defined enclosure of some place upon the architect. An ancient loop, as round as an orchestra or a lecture hall: society fabricates these motor temples which fabricate society.

These localities in fact vary little from the Athenian Pnyx, into which the Assembly of the people was packed, crammed together, between the Areopagus and the Acropolis, and some contemporary Parliament. Vocal carrying-distance has scarcely changed in three millennia. Yet, the way we have made evolution set sail from bodies towards technological objects, the microphone, first, then other media, whose connections extend to millions, even billions of people, make these formerly closed places explode towards an unexpected global. The Voice is no longer crying out in the desert but rather towards an overpopulated absence. A new loop and, for once, one of hominescence: society – but do we know which? – fabricated these technologies whose carrying-distance exceeds it in order to touch the entire human community. Will these technologies fabricate it? In any case, this new stroke of hominescence will force me to redefine a humanism tomorrow.

Of course, in the interval, the book had succeeded the voice as social tool since books reached populations the voice couldn't attain. Hence the vast effects of writing, from its invention and after printed material had been developed, on law and politics, economics and religion, the sciences and education. Victor Hugo had written this, in front of *Notre-Dame de Paris*: 'this will kill that'; the book will kill the edifice. For buildings organize the voice, which the written propagates to the ends of the earth. This prediction (literally prophetic, although false since no medium kills its predecessors: as evidence, precisely those enduring buildings) is being repeated today by the philosopher: this goes beyond that; the universal social tool – and we are witnessing its birth – adds up and surpasses the entirety of the performances of its predecessors.

Quasi-objects, world-objects, society machines

Let quasi-object be the name of that token that runs from hand to hand, that ring that passes along the stretched string in a circle of children singing 'the ferret of the woods, my ladies' or the ball with which two teams play; this peace pipe, these coins … mark out the fluctuating and regulated relations that become established or can vanish between the subjects of a collective.[1] Here already are little social tools; by running from palm to palm, they contribute locally, from neighbour to neighbour, to constituting a group. From sports spectacles to the theatre, we thus move from balls to words and from this token to meaning. But never did we dispose of a tool, nay, of a machine suitable for touching, no longer a few individuals, but a vast population at a stroke. To make the global mass gel or listen, tyrants and televisions dispose of violence, propagating lightning-fast. Had it already been remarked that violence could simply be taken to be a social tool?

Let world-object, as I recall, be the name of the atomic bomb, nuclear waste or the internet itself because one of their physical dimensions – energy, time or space – attains the scale of one of the dimensions of the world. Never did we build any tool or machine suitable for working on the global world. Yet today we can make the hole in the ozone layer vary and spread enough gaseous waste to slowly warm the planet. When these world-objects quit the strict status of being tools or machines suitable for working on things so as to become quasi-objects, like the media in general, they transform into society machines.

Quasi-object plus world-object equals social tool: carried by the UMTS, the internet really occupies the space of the world and virtually touches every human. The integration effect doesn't merely concern inert things, but also and at the same time social actors. 'We' now build machines suitable for working on the world, but also on humanity as such, 'us': humanity-quasi-object. The loop I have just outlined becomes, globally, a set of machines, of *logos* technologies, of techno-logies. Have we drawn the consequences of these new scales? I dream of a word and a verb that would replace, in such cases, the verb 'to work on' as well as the substantive 'machine'; defined for other ends, the old terms lead us into error.

The thing itself and consensus

So truth is established by the evidence of the thing in itself and by the opinion of the greatest number or of selected experts. We hesitate to call truth universal or necessary, so much are the two linked. This reminder brings us back to the new society machines; let's read, listen to, watch on the television, for example, some report on a crime, a war, a scandal, a catastrophe; few of them are on peace, virtue, justice or happiness. We see, self-evidently, the persons implicated, the landscapes burnt to a cinder under the volcanic lava, the ruins of the city ravaged by the cyclone or the bombardment. We see the thing itself. The best truth therefore shines forth to the eyes since the evidence shown there concerns neither opinion nor consensus.

Yet no one would ever broadcast such images if they didn't have any chance of capturing the attention of a large audience. Their probability of being present at the screen increases or decreases according to the machine for measuring this audience, called Audimat in France. Likewise, a website is worth the number of visits it receives. So it's a question of messages-numbers, of images-audience, of precisely what I am calling quasi-objects. The more we become interested in signs, the less we see the things themselves they show and the more we follow the tokens passed, here virtually, by the people making up this vast and countable audience. In other words, we see the Audimat through the intermediary of this crime, of the scandal, of the war or the catastrophe. We contemplate less a thing than the cause that glues [*colle*] us together. We see that on which the collective is focused in general.

Everyone knows – the learned since Aristotle and the others since the most ancient myths recounted horrors worthy of the Grand Guignol – that the collective doesn't care one whit about peace, without savour, about virtue, odourless, about naive kindness, about justice, too reasonable, or about happiness, foolishly abloom, but adores violence, terror and pity, as Aristotle said, the guilty party and the victim, the killer and the dispenser of justice, when all's said and done, corpses. We therefore quite simply and madly repetitively see what we have preferred to see ever since we have lived together, the archaic glue of the collective. So the media unceasingly repeats the same news, the most ancient that can be: crimes, wars, catastrophes, deaths, killers and bodies, those responsible for denouncing or for lynching. We see this contract; we hear it being set forth repeatedly in a deadly dismal manner. I turn on the TV to see this social contract virtually in real time. By switching the set back on again, I sign and re-sign this contract. Rousseau never dreamed that, materialized in machines, his *Social Contract* would one day appear in the light of the concrete. Yet our object functions like a virtual and global polling station, ceaselessly fed with the same quasi-objects of unanimity – I was going to say of the general will. No one will be surprised, not at the fact that the media behave as though they had read Aristotle more than Rousseau, but at the fact that their messages come out of the most archaic attractive well of said will: murders and dead bodies. Nothing glues us together better, since the foundation of the world, than human sacrifice. Every image boils down to this, the old cement of the collective, the old machine for fabricating the thousand glorious gods of polytheism.

Another – lateral – loop: beneath the outsides of the thing itself, it's less a question of the thing than of consensus. Violence has virtues that are stronger than speech: universal and without any need of translation, it touches all of us before meaning and without it.

Myths and beginnings

As during any period of transition, in which technologies move by sail or steam, we are living through that fascinating time in which the new machines still function with the old motors they could easily do without. For, to glue the collective together, we formerly and recently had need of violence because it spreads, because it optimizes relations with regard to their speed and transitivity, because it propagates itself the most easily in the world; a diabolical causal loop increasingly feeding into itself, violence is born from violence and never ceases being reborn from itself. This almost automatic analytical extension allowed a quasi-object, ordinary and local, for which battle was kindled and raged, to ensure a certain integration.

Our new technologies end up directly at the result – fabricating unanimity – without having any need for the ancient glue, that violence done to man by man. Such technologies (radio waves, optical fibre, pixels) are sufficient for this. Everything happens as though we knew how to and could reach a goal and as though we were so afraid of missing this goal that we used, just in case, the ancient formulas, dating from the times we didn't dispose of these machines. Steam allowed us to speed along at twenty-five knots on the lovely sea, but in the event of wind, we in addition unfurled jib and brigantine to reap a half a knot more. We enjoy effective and recent technologies; we fill them with primitive myths. So can't we eliminate them? Must we detect intention or stupidity here? Neither of the two – heaven forbid I should denounce – but simply the usual state of transition ages, the normal look of phase changes. For the new machines directly address the large number, the collective. As a result, the collective reconstructs itself and does this as though it were beginning to do so. We are living through this event in its nascent state. For the new technologies favour social newness, and this latter affects the new technologies. The fact that this nascent starting moment favours the re-emergence of the most ancient myths, of grand narratives of beginning, what could be less unexpected? How, tomorrow, are we to quit these old habits, to learn to do without sails, to abandon obscene violence?

Myths, the hand and the tool

Abandoning theory and observation here, sociology becomes practice, technological and experimental. Every new pilot of these social machines says to the collective 'go', and it goes; 'come', and it comes; 'stay with us', and it stays; 'pay', it pays; 'give', it gives; 'be quiet', it keeps quiet; I say the just and the true, and it believes it or at least acts like it believed it. So these are society tools, like tools for wood or iron, for these machines 'work', in fact, on the social bond. Politics therefore finds itself changed by this because these technologies take an active power over social phenomena and take precedence over men of state; law and cults find themselves, in their turn, changed by this since they fabricate victims and gods. What philosopher, even a recent one, would have believed that one day machines suitable for fabricating politics, morality and religion might exist? That *Homo faber* would shape *Homo politicus*? That the transmitter would take up all places?

How do politics and the institutional, for example, command the collective? This verb of behaviour refers to a hand, as I have said: but which one and whose? We have never learned, since Adam Smith spoke of its invisibility; Plato likewise recounted a king Gyges disappearing in front of his blinded subjects thanks to a ring he would slip on his finger, thus prefiguring Smith's invisible hand. On it and its fingers an iron glove sometimes appears, visible and tangible. Power terrorizes the social in order to govern it. A bad sign. Whoever displays this force, whether police or military force, confesses that he is ignorant of why or how he governs since he only knows how to resort to death, whether violent or legal, to secure his place. The invisibility of the hand that orders the market, the invisibility of the finger that rules while wearing the ring, even the invisibility of the king's body betray an essential ignorance of the societal, a kind of collective unknown made perceptible, if I may, by this invisibility. Language, sometimes, music, as well, the charisma of this or that person stirred these dark masses, but above all violence and death. This is one of the great factories of human myths, that black pit without light on which institutions are built. Power sits on a throne that hides this hole of ignorance.

When I try to speak to three thousand people, I only see, quasi-blind, a narrow human horizon. My voice, not strong enough, barely reaches them. Let it make use of a mike: this is the first social tool for which the hand that holds it finally becomes visible. So hand and voice negotiate the reactions of the crowd. Radio broadcasts, a screen, the internet next reach what has remained inaccessible ever since we became humans, reach this black hole in which large populations bustle, a black hole from which violence, power, rites and myths irrepressibly shoot up. How are we to master such scales, such numbers from which gigantic energies flow? Yes, these new technologies still today remain in the grip of myths: they fabricate myths at the same time as they do gods and victims and thus take us back to the beginnings of our narratives, to the beginnings of the construction of social phenomena, in which we invented myths and politics in the absence of any machine suitable for working on the collective, a fortiori in the absence of any hand tool, and, without knowing it of course, owing to this absence.

We now dispose of these tools, but we don't yet know their technology, in the ancient sense of this term: the way to make use of them. So let's not denounce their foolish apotheoses or this daily fabrication of gods, the savagery of crimes incessantly represented, the crude brutality of their culture, their very culture of perennial denunciation, for all of these practices simply betray their immersion in the origin; this origin is indeed taking place since, for the first time, the working hand is showing itself and working on social phenomena, reconstructing them and informing them in the most traditional sense. For lack of a new political philosophy, we are still tangled up in this limbo, stuck to these beginnings. Nietzsche cried out that the West didn't know how to create myths anymore; what an irrelevant consideration for the present! Visible to the point of blinding glare, our myths even let us see, through their constant violence, the very origin of all myth.

15 THE OTHERS AND THE DEATH OF THE *EGO*

But are we recreating autonomous people who know how to say: I? The more contemporaries employ this word, love it and flatter it in that Latin no one understands anymore, the less they hear it in its origin sense: some person, they say, endowed with a big fat *ego* fills space up with his voice, with his obesity of soul or obesity of body sometimes; reputed to be equipped with the same organ, some other person, political, manipulates relationships and people to bend them as he wishes. In these two examples, the 'I' designates the 'we' since it's a question of glory and theatre, of exhibition and power. Without the others, this poor *ego* deflates.

Conversely, St. Augustine's inward subject, intoxicated with God, Montaigne's subject, amiable and fluid, Descartes's subject, reflexive, or Rousseau's subject, self-complacent, 'selves' Pascal, in total, judged or would have judged to be hateful, the subjects dealt with in the confession manuals striving towards the salvation of the personal soul, all of these *egos* of the Christian tradition and of the philosophies of the Latin language withdraw into prayer, into their stove-warmed room, their library, into Port-Royal or on to St. Peter's Island, in brief, shut themselves up in a solitary retreat so as to meditate in isolation after the example of the anchorites, de Rancé, for example, a socializing courtier become a Carthusian monk. Social space becomes the *non-ego*, not only buzzing with vanity outside the cloister, but, in addition, reputed to be dangerous for the development of the person.

As fragile as the self, the great works of our culture were born from this *ego* cultivated inside a maintained enclosure. Either, empty and vain, the *ego* swells up from public acclaim, or it lives from avoiding it. These are two contradictory 'selves'.

Truth

Taking refuge in a sort of cell to avoid the sounds and furies of war, officer Descartes, in silence and through solitude, discovers his *ego* as the subject of truth. It lies not only at the foundation of thought, but of existence itself and of true statements. This discovery takes place in a meditation in the form of a three-way dialogue between the 'self', precisely, the evil genius, a diabolical tempter, and God, the guarantor of this true in the final analysis since the creator of eternal truths.

As fragile as this self, the big truths of our sciences were born inside analogous protections. Social space didn't, as far as I know, favour Galileo, condemned, or Descartes, taken refuge in Holland, or Spinoza, stoned, or Mendel, forgotten, or Boltzmann, dead by his own hand in Trieste on the shores of the Adriatic, or Majorana, vanished, or Turing, driven to suicide by British laws after having saved his country. Four-fifths of Nobel Prize winners today receive this recompense, in the end and at end of their lives, after other juries gathering their peers rejected their research projects; first of all condemned, lastly awarded prizes. These unfortunate events can probably be dated back to an era, perhaps bygone, in which the subject of the sciences was identical to the one described by the humanities since no one could establish any difference between the humanities and the sciences; but everyone saw the difference that separated public action from private life.

But truth today, quite the contrary, is born, they say, from open and fierce debates in which, as in the stadium, the theatre or on the fields of battle, dialectical confrontation ends with the victory of one party and the defeat of the others. So the foundation of the true quits the meditative solitude of the self, protected from social noise by high walls, so as to, quite the contrary, rejoin this very furore and the laurels of triumph. Before the invention of physics by the Presocratics, the most ancient truth, archaic Greece's *alétheia*, as we will remember, also meant the social glory acquired by the exploits of those who were lucky enough to find a famous poet to celebrate them: it became confused with honours. By facing trials of all types and condemnations, whose echo is repeated by Galileo's condemnation, the Presocratics discovered physics by defining a truth the opposite of that truth, that is, one independent of all debate and even stopping them from buzzing.

Debate, where is your victory?

Vitalism's quarrel with mechanism ended, in the last century, with the defeat of the former. Therefore mechanism triumphed. So they say! For machines like Turing's, which today we know serve the living in their reproduction and development, differ so much from the machines Descartes or La Mettrie used that it is astonishing they bear the same name. The triumphant mechanism, which crossed swords with vitalism, has vanished just as much as the defeated. The squabble between Geoffroy Saint-Hilaire and Cuvier ended to the advantage of the latter. But ever since biochemists located the homeobox, Geoffroy has reawakened in quite good health: the general body plan of organisms has returned. Pasteur's dispute with Pouchet ended with the defeat of the latter. Who today would dare to speak of spontaneous generation? Answer: everyone, for biochemists are precisely searching for how the first DNA was formed from chemical elements that were indeed a little metallic. Yes, every living thing comes from living things except this acid or an RNA, which constitute the living itself. Semmelweis's blacklisting by his fellows from every country in Europe ended with his death in despair. A few years later, Pasteur showed that he was right to make obstetricians wash their hands and that the victors were behaving like filthy criminals. Do we need to make this tragicomic list longer?

Either the true is born and then grows with public acclaim, or it develops by avoiding its noise. These two truths seem contradictory. They can probably be reconciled because they follow one another in time. The science of isolated people, who are intuitive and rare, is followed by the science of professionals, who are gathered by the dozens in as many laboratories scattered across the world and researching the same thing. The original subject, as the foundation of the true, is followed by an interactive we. So who built the thermonuclear bomb, who discovered this or that subatomic particle, who will have deciphered the human genome? The collective subject of a truth so common that any equivalent group, elsewhere and at another time, would also have encountered it. So the number of those who sign it changes; our networks multiply these instances and these signed initials. But one cannot straight out say that the multiplicity of thinkers, linked by an internet that would resemble a big brain, constitutes a more powerful single thinker: fifty policemen marching in step don't go fifty times faster than a single walking policeman; on a long hike, there's even a large chance they'd jostle one another and therefore stop. In order to improve the guarantee of truth, entirely different conditions of verification will be needed; and even if they existed, the truth could still escape them.

Existence

But before truth, existence itself is at stake. I think, therefore I exist. What philosophy doesn't teach – to the point that it has to do here with an exceptional agreement between almost every school, nonetheless so opposed on every other terrain – that social life flies by means of empty appearances, that it drags after itself masks and lies, vague shadows and power without any future? Even the long list of the masters of the discipline that became counsellors of kings, from Plato on up to Seneca and from Voltaire on up to certain of my contemporaries, in the end turned away in disgust. As though public life irremediably tore up existence, as though solidarity itself vanished there, as though we could only build enduring relations with others on the prior condition of constructing oneself outside the vanity fair. The truest relation to the other, the entreaty of love, develops and lives in private. Sociology has a theatre as its object.

Unlike this certainty, the contemporary *ego* is constructed only in and through the relations formerly called worldly, relations outside the monastic enclosure or the windowless monads. The former loss turns into gain, as though others alone could ensure the self its authentic existence, its growth and development. The multiplication of the means of communication no doubt contributes to the new order of things so that a starred self at the centre of a network, a crossroads that's all the denser the more numerous the roads that depart from and lead to it, becomes opposed to the first person of the *cogito ergo sum*. While for Leibniz, for example, monadic solitude sought without finding them relations that only God could ensure, today relation precedes existence, ensures and founds it, unfurls and enriches it. Either the body closes itself up inside a Carthusian monastery, or even stretches itself out on the sole of a flat-bottomed boat like Rousseau tossed about at the mercy of the winds on Lake Bienne in order to feel his own and original existence, or 'my thousand-voiced soul, which the God I honour put at the center of everything like a sonorous echo' resounds with relations.[1] Here are two existences that seem contradictory, but one of which no doubt, again, follows the other, as though Echo the nymph had replaced Narcissus or hearing sight.

The I lives thanks to idle time. But who today wouldn't complain about the boredom of a time without any diversion? When Blaise Pascal described the room in which one must know how to remain so as not to expose oneself to the misfortunes of the world, he was talking about the private life, that idle time from which arose, in boredom, sometimes, and silence, that little noise lacking in strength sometimes called the voice of conscience.[2] Where am I to hear this frail murmuring if the music never stops, if my room, if my private space, if my home, formerly cut off, transform into a public square equipped with radios, telephones and email, with televisions and faxes, with that UMTS I've called the social universal? Outside life enters there to the point that it destroys the home and private space. But the I is only constructed in this room, now become a marketplace shot through with the collective's noise, with political furore, authoritarian publicity and obligatory communication. The I dies from the us.

On syntax and the subjunctive

Remember a language written and spoken by altruistic, refined and sensitive women, Marguerite de Navarre, Madame de La Fayette or Madame de Sévigné, the Portuguese Nun ..., or by egotistical men, *Adolphe, Dominique* ..., both endowed with this 'I'.[3] Now unlistenable, even unreadable, their style was packed with a subtle syntax whose verbs of intention would launch, after the subordinating conjunctions, the subjunctive and sometimes its imperfect. Become rare, this tense said the subjective.[4] Expressions whose multiplicity unfolded the plastic landscape of inwardness wove a text whose bifurcations hugged the contours of naive clumsy turns of phrase and mendacious fluidities of style, of good and bad faith, of theatrical or sublime love affairs, a landscape as differentiated as the ascent of Mount Carmel. The subjunctive–subjective stretched the soul into the length, width and height of a waiting time we now call idle time. 'Real time' or the indicative reduces this volume to nothing. The suppression of distances has just tossed this slow waiting on to the side of mental illnesses and suppresses the memory of an era when attention sculpted the soul with its folds. Our language has just lost its syntax and, as a result, its Maps of the Tender, the dimensions, the relief, the volume of a self enriched by the topography of the you. We communicate, but in abbreviated messages. I am highlighting here the subject of the verb, this we that becomes a substitution for the old blunt-pointed I. We are no longer subject to the subjunctive.

Relation precedes being. This indeed is the saying of my philosophy; I have never spoken about anything other than communication; never have I described consciousness or desired to penetrate arcana for which I didn't possess the key, and those who did claim to have it seemed to me to be charlatans. I resign myself to considering the soul to be a plastic virtuality that flows in language; I therefore stand by what I said but measure what is lacking: by multiplying networks, by restraining messages to their skeletons, by replacing the *ego* with the echo of the we, the biological evolution of the species runs the risk of leading us back to societies of insects without any neighbour, termites, ants or bees, all the more so since demographic growth, the globalization of the economic, and artificial reproduction are accelerating the process. Some central bank of ova and sperm will take the place of the queen, and we will work as specialists for the mechanical continuation of the universal network. By what dynamic are we to rectify this possible fall?

My language doesn't reduce to an instrument of communication. Who in fact am I before saying to myself that I am? Who am I without this inward speech which, tangent to silence, constructs me, even if, as a child, I borrowed it from the woman who characterizes my maternal language? Who am I if not an emergence whose noise rises above a black and silent arch, from which a sustained murmuring lifts itself above this background noise, from which some first music, rhythm and song, shoots up from this murmuring, from which the first stammering rises above this music? Who am I if not this house of cards, each fragile panel of which is said in order to surround or protect this arch or silent and black arcanum? My entire life I've remained silent and have only discoursed in order to conceal this silence, with my books constructing a suspended vault above this cavern.

Happiness

The porch of my friend who died last July is decorated with a moderately coloured mosaic on which you can read, still in the same dead language: *Beata solitudo, sola beatitudo* ('Blessed solitude, only blessedness'). Living in seclusion, despairing of his experiences with the collective, he finished his life the way La Fontaine ended the books of his *Fables*, by extolling the private life, finally appraised to be the supreme good. In my early youth, the people themselves repeated proverbs about the happiness of monks and philosophers, considered by them to be self-sufficient, above common needs, therefore effectively happy. Today, quite the contrary, we proclaim the solitary unhappiness of those who lose all relation. We consider them to be excluded, sometimes sick, in any case in despair.

Just as today administration forces its subjects to live administratively, so everything happens as though the collective had succeeded in making a solitary and secluded life impossible. At the very moment politics is no longer seizing the forces that work on and transform the collective, private life alone is becoming entirely political. Society and the public impel us to live socially and publicly; networks of relation make life without relation impossible; truths acquired in common exclude the truths whose intuition suddenly strikes the isolated researcher; ecstasy of the body or of the monad-soul is taken for being sick. In sum, for the true, the existent and the happy, the *ego* has just died. In citing it, we are no longer talking about anything other than a shadow witness to a vanished culture. With Descartes dead, it remains for us to write: I link myself, therefore I am. Relation precedes all existence.

Montaigne's *ego* and mine

Once again, this doesn't date from yesterday: believe Montaigne in what he does more than in what he says. When you hear him confess that he is only talking about himself, smile and move on. For he cites pages full of Plutarch, Tacitus, Lucretius, Virgil, Epicurus and Socrates ..., his neighbours, the Parisians, the inhabitants of Bordeaux, but also and above all American Indians. Listen to him well: as he is recounting Alexander's conquests or Diogenes's barrel, he is captain or beggar, proud, chaste, famished, mocking. His thousand-voiced soul, which the God he honours put in the centre of everything like a sonorous echo, was never the soul, stable, of that vineyard notable who strove after the mayoralty of the port. No. Like you and me, if I may speak so, Montaigne enjoyed a mixed, iridescent, striped, constellated soul, reflecting without cease the austere, the gourmand and the lascivious, the solitary and the solidary, the shy and the jovial, the vagabond and the homebody, the pious and the atheist. I have never understood how one could *not* be at the same time and in all respects the entirety of humans that can be encountered and have been encountered. No, I am not me, like a point, and fixed, but the cloud of people possibly close to me. Flowing, temporal, diverse, my identity has nothing to do with the ontology of being or with the spatial, exclusive, single principle of identity, but rather with the possible. Yes, relation precedes being; I am my neighbour.

The current coexistence of two types of *ego*

Again it's a question of evolution in the biological sense of the term or of a quite precise stage in the process of hominescence. Two periods succeed each other, one of which no doubt began with Christianity, a religion proposing the salvation of the individual soul and therefore in which the faith of the person himself or herself is demanded, a person ignored by the previous religions, wholly founded on the city or the people, in which therefore a new agency becomes involved, quite precisely that *ego* the

Greeks and the Latins only knew vaguely since the famous 'know thyself' only invites an assessment of its bounds. Its credo begins with this unwritten pronoun: (*ego*) *credo*. Christianity at the same time quits the collective glued together by the sacred and abandons as well the enchantment of the world, as Auguste Comte demonstrated, but it above all abandons the religious that functions as a collective bond so as to address itself, from the start, to the solitary soul, faced with its fate and its eternal salvation.

Consequently, the contemporary death of the *I* strikes a terrible blow not only to European culture, educated in Montaigne, in *Dominique*, but also to Christianity, even if one pretends to believe that it is rejoining today the private sphere: no, it had invented it. Religions are returning to the archaic state of social cement, or rather and conversely, the new and powerful collective glue is expanding into fundamentalisms, hence the wars that set them in opposition, as during the most ancient times. Too modern in the face of this strange regression, Christianity is becoming difficult to understand again. Since it addresses itself to the personal soul and to its fate beyond time, it gives a devastating critique of every fundamentalism, understood as social concrete. The end of the *I* lastly inflicts a potentially mortal wound to a culture whose universalism owes its influence and its creativity, for worse and for better, all the way to the rights of the person, to this existential agency.

Rare men and women who still have private *egos* coexist with contemporaries, in droves, having public *egos*; how would these people who are organized differently or bear an organ absent in their neighbours understand each other? In evolutionary terms: Which of them will eliminate the others, as during the days when *Homo sapiens* coexisted with the Neanderthal? Therefore wager on the best adapted: Should you bet on the one who arms himself in order to win in every circumstance or the one who profits from a refuge in the event of shipwreck? The first one was born under the conditions of peace that have prevailed for merely a half century; he loves open war for never having suffered it. But will he always win? How will he who seeks to win survive in the event of defeat? Since life numbers more obstacles than non-constraints, more sufferings than perfect joys, more failures than triumphs, more spurned loves than crowned ones, I bet in favour of a reconstruction of the I. But how can we create a vault of silence in the universal reign of noise?

When the intuition comes, such a silence is necessary to hear its gentle breeze that Death Valley's lowest low spot, that the Hoggar Mountains' high red rock, subtly vibrating under the noontide sun, that the Kalahari, the Gobi or the Atacama Deserts still ring with too much racket for the listener to receive this intuition; the slightest rustling drives it away. This is the mute secret of creation, artistic creation in particular, of listening, period, of waiting, of attention to others, of inwardness. The silence and gentleness of this undecipherably enigmatic modesty allows survival in a world devoted to the perpetual background noise of communication, and which risks constructing collectiveness by destroying what we were calling the person.

Just as Descartes, in his stove-warmed room, doubted, so *I* often unplug. Education by means of the new technologies is therefore complemented with a pedagogy of disconnection and an ethics of detachment. The future belongs to the contemplative orders. We shall be saved from the evolutionary fall towards insect societies by the one who will invent a new generation of monasteries: this word signifies a paradoxical association of solitaries and solidaries. We shall have need of a Saint Benedict, of a new self and of different neighbours.

PEACE

The passage to history

Three gods, Jupiter, Mars and Quirinus, rule over traditional Indo-European societies, which consist of three corresponding groups devoted respectively to religious rites, war and production. Classified in this way by Georges Dumézil, this ancient society is found in the Middle Ages and lasts all the way up to a recent era. What date did this long tradition disappear?

I have told of the end of Quirinus, the protector of ploughmen; not the protector of agriculture of course, but of a certain cultural model tied to the ancestral domestication of species of flora and fauna. Many people seem to hope for the protection of Jupiter, the first among priests, the only protection I doubt, for the proliferation of signs and of communication can only relaunch processes of the religious type, like monotheism at the beginning of writing and the Reformation, the child of printing; since, for the first time since Eve and Adam, life and the ways to reproduce are changing, our symbols are changing.

Trembling delightfully with hope but seized by the fear of wandering, I shall describe here the death of Mars, military ruffian, soldier, warrior, the god of death by weapons.

The loop of war

For I shall soon live on to be among the last to know, through the body's memory, that war occupied the normal horizon of every social group. My culture dates from the fall of Troy, from the murder of Abel and from the Passion of Jesus Christ surrounded by Roman army rabble. My short life, as for it, woke up from the Spanish Revolution to the world conflict of

1939–1945 and grew up from the Indochina War to the Algerian War. Living amid the corpses, I experienced peace quite late.

Political leaders traverse History bandaging the wounds the previous war cruelly inflicted on their countries and preparing the next one in haste, inevitable; they didn't experience any cessation in this perpetual series. Reciprocally, they only led empires or democracies by relying on the threat, imminent and constant, of conflicts. Lastly, they only existed due to this inexorable swinging. The state was born from war and is always strengthened by it. At the outset, it acquired its wealth through wars of plunder and by killing those who possessed it in order to pillage it. Having demanded taking. All assets came from theft and theft from violence. The state, say historians, invented taxation to gather the means to wage war; these levies remained in place during the extremely rare times of peace. From the elite to the animal – frigate bird or rat, parasite or predator – *Homo terminator* behaved towards its fellows as every living thing does with every living thing, a rare beast nonetheless for delighting in intraspecific murder.

War creates the state, which creates war, which creates History, which creates war, which creates man, which creates war – this is the catechism in as many spiralling loops taught by almost all our philosophies, armoured from fights to the death, combats, brawls and debates considered not only to be normal, but to be the exclusive motors of renewal and knowledge. Without war, there can be no state, no history, even less can there be man, inventions or advances. From economic competition to competitive sports, cultural formations, philosophy included, are all taken to be wars continued by other means. This bloody civilization looks down on projects of perpetual peace, calling them utopian dreams; squabbling takes on conceptual loftiness and dignity, and peace takes on the ignominy of naive idealism. No one does or writes History with good intentions. Never search for the causes or the reason for a war; war replaces causes and reasons.

And what if the situation were reversed?

During recent decades, I often dreamed that the generation following me would differ from mine due to the experience, whether lived through or not, of cannons and bombings, sirens, famine, corpses in the trenches, blood and guts in the mud. For all of a sudden violent death and its bereavements had quit the world view, the usual behaviour and ethics of my students. War impelled me from behind and dictated my previsions; my students weren't reading this alpha or this omega anymore, the current letters of the time. What will become of them, I said to myself, when the next war takes them by surprise and without being prepared? Will they be able to survive its extreme conditions? The break between the time of Reagan, Mitterrand or Kohl, former combatants, even in salons, and the age of Blair, Schröder, Clinton or Jospin, deprived, for their part, of such enduring memories, is recognized by the fact that the new generations who have just followed in the calm of peace, without any other knowledge of war than by a few testimonies, books or movies, make so many errors regarding the years we nearly died and which we don't talk about for this reason; we never recognize our memory in their histories.

This absence of world conflict only concerns the countries said to be developed, capable, because of their power, their sciences and their high and noble culture, of unleashing violence elsewhere than their homes, on the totality of the planet, at least by means of selling arms. For them and between them, this truce has lasted in the West since 1945, not taking into account the sabre-rattling of a war said to be cold, in which the confrontation of two gigantically armed giants ended in the implosion of one of them, while waiting for the implosion of the other, and the peaceful fall of the Berlin Wall. Everywhere else, a thousand conflicts, often cynically kept going by the very people who enjoy their distance, the spectacle they provide and the clear conscience they produce, deploy their multi-millennial logic and its atrocities. Even limited in this way, this recent peace occurs as an extraordinary exception. When has a certain percentage of humanity enjoyed such a long interval of time without any major wars? Never. Soon it'll be sixty years of calm on these fronts said to be developed, whereas in past centuries, the seventeenth for example, only seven peaceful years could be counted in Europe. I doubt that this ultra-paradoxical moment can be compared to the Pax Romana or the languors that fooled certain pharaohs and Chinese emperors, tyrants who held, in all three cases, kingdoms in which hierarchy necessitated cutting heads off in mass. Are we living in the midst of extinct volcanoes? What equivalent of plate tectonics can explain this singularity?

When Mars goes absent, what consequences are there?

Are we moving towards the end of the warrior? Our career military men now wage fewer wars on foreign powers and instead wear themselves out trying to prevent others from waging them; they have to practise the reverse of their profession. Public battle reverses as well, and in the opposite heading, the advantages it used to procure. Our world now lives total victory – in which the triumphant army tallies twenty deaths by accidents and the losing army two hundred thousand under bombs – as a massacre that ends up covering the winner with the blood of its victims, even if the war seemed 'just' at the outset. The dialectic is reversed: we are no longer living the era of victors but rather the era of victims. Who can deny the progress, at least on this point?

If Western philosophers had been consistent with the traditions they commented on and respected, they would have asked themselves whether, in the absence of the war motor, humans still existed worthy of this name in their tranquil countries and whether they had invented anything whatsoever there during this strange interval. Yet never before had science advanced, during those few decades, at the pace these countries experienced. A counterexample to dialectics of every stripe, might peace contain a drive motor? Furthermore, during the same blessed interval of a half century, and despite the obsessive fears of the Cold War and the blunders of decolonization, the countries concerned had produced immense wealth. Lastly, never before had the Earth been covered with as many humans; the demographic explosion accompanied this peace despite the multiplicity of local conflicts.

More than Kant, Pulcinella himself was therefore in possession of the secret of the vast abundance issuing from 'perpetual peace': it required not destroying.[1] Is this sufficient?

The first objective cause: The bomb

So where did such a long peace arise from? No doubt from the risk of holocaust, in the etymological and religious sense of the burning of the entire human race by fire, quickly read in the Hiroshima explosion and the various thermonuclear bombs tests. While war imposed itself as the unsurpassable horizon of historical society, the energy power of these weapons suddenly filled the width of that horizon with its glare. Surpassing with its strength and its effects the global entirety of conceivable conflicts, it even surpasses their sum. By totalizing the polemical time of the old history in this way, it stops its motors. Consequently, this diabolical invention brought about a result opposite to the one for which the scientists and politicians conceived and developed it, often experimented with and used it, twice only. Through an excess beyond these integrations, it wins at a stroke every possible war. So the secret to peace lies in total eradication, the disappearance of the human race and of its planet, perhaps, a secret seen by everyone as the now accessible horizon of every world conflict. *Si vis pacem perpetuam, para bellum totale* ('if you want perpetual peace, prepare for total war'), better, for possible complete destruction.

I often have doubts about such a cause, which is cited everywhere. Our archaic and tragic culture, brought about by war and its deaths, maybe leads us to give it credence. Whether peace comes from war, whether life is born from death, whether invention emerges from destruction, or perpetual peace arises from maximal violence, the same maniacal catechism always returns. This catechism steers the warrior's arms, draws the politician's public square, inclines the historian's interpretation and colours the philosopher's reason. Might peace be established, on the contrary, if we forgot what must indeed be called an ideology, one rooted in practices common to every living being? Are we, in these brief decades, terminating our long career as *terminator*, whether as individuals or collectives? Are we going to change not only our idea or ideology but our conception of life? Is life itself transforming? Yes.

Hominescent causes

For this first cause refers to the beginning of this book, where I name it one of the new deaths. For never before had humanity envisioned its own eradication; worse, never before had it planned or programmed it. Faced with this unknown threshold, a sudden stop. Yes, this event closes a long epoch, precisely the epoch of hominization that can be considered to have begun with the awareness of individual death. Changing deaths, changing times: so the closure of this epoch launched us towards another epoch.

It happens that the new epoch lived still other newnesses of the same span and the same depth: the newness of bodies, of the relation to oneself, to the land, to the world and to others. So when life expectancy lengthens to the point of doubling, will the potential hero so easily accept laying waste to this period of time he could still enjoy? When, during this long life, pain lessens, will he wish for it? Will he not consider all war to be a collective pathology to be eradicated with as much vigour as small pox? When the worked plot of land loses its economic value to the point that agricultural production falls to below 5 per cent of the gross national product, will he defend fields and harvests to the point of sacrificing for them an existence that has become so precious? When, by means of world-objects, house walls stretch towards a universal horizon, will he be prepared to battle to the death *pro aris et focis*, for domestic altars and hearths, particularly as domestication is dying out? When borders become erased and every morning he who says 'bird' communicates with those who say *Vogel*, *oiseau, uccelo* or *pássaro*, will he hate them as intensely because they don't speak the same language or pray to the same gods?

Thus all the elements of this time of hominescence are united to form an unexpected hero who will be made sick by every armed conflict and who will fight only to abolish every death penalty. This new woman, this new man of peace are born from the chapters of this book.

The second bomb

In the deserts of New Mexico, towards Los Alamos, the Manhattan Project assembled, as we know, the elite physicists of the day and gave even more glory to those who sought to become and indeed did become global divas of knowledge. Here, the stardom of the new geniuses was prepared at the same time as a technological crime against humanity, cruel to the scientists' consciences. Thus, starting from Hiroshima, the relations between the sciences and society were no longer as good as before.

Additionally, between the years 1930 and 1960, a second bomb exploded over a long period, a bomb owing nothing to the hard sciences but everything to the organization of societies, their governments and ideologies. Thus, the century experienced two slaughters, a technological one and a political one, in parallel. In Nazi Germany, in fascist Italy, in Francoist Spain, in Soviet Russia, in Maoist China, in Cambodia ..., totalitarian regimes killed, by the tens of millions, more people than the scientific crime. The end of the Second World War didn't entirely stop their effects.

For these regimes affected Western consciousness infinitely more than the scientific crime did, as though a collective guilt suddenly emerged at the memory of horrors for which groups as such thought themselves to be responsible, inevitably. Assuredly, there wasn't anything we could do about the energy equation, nor about the speed of light, whereas we could have fought with more determination against our social ills, whose development depends on our behaviour and general will. Let the Roman soldiers of antiquity leave behind them no Sardinian inhabitant alive, let Julius Caesar, in Gaul, murder every adult, women and men, except for a few old people and children forgotten in the woods and ditches, to the extent that we have lost the language of those we call our forebears, let Louis IX completely eradicate the Cathars from South West France to the extent that my family only retains a vague cantilena from them, let English generals, by massacring Indians, Hottentots and Zulus by the tens of thousands, conquer an ignoble glory, intensified by an Australian pre-Gulag, and we tolerated it, distracted by their distance in space and time, I mean by forgetfulness and racism ..., but let a neighbouring country, installed at the head of Western culture and science, to the point of arousing the admiration of its competitors, plunge in a few weeks into the Shoah, let philosophers, at least in my country, continuously support three or four of these regimes for more than thirty years, from Stalin to Mao and Pol Pot, will we ever recover from such an epidemic?

Since the word 'religion' signifies in our languages the set of bonds that assemble a collective, we are all of us today glued together by this haunting memory. Beneath the external appearances of atheism, we are silently practising a lay religion whose constant thought resembles the thought of original sin: ineradicable, violence follows us step for step and never leaves us whether we're practising it or regretting it – what can we do to negotiate its ravages? Neither the advanced sciences nor lofty musics, nor philosophies presumed to be profound preserve us from it. That never again! Had we ever experienced such a guilt? I hear it said everywhere that guilt as such weighs on us more heavily than the crime or the sin and that it must be eradicated. Who, however, among their victims, wouldn't have preferred that Julius Caesar, Napoleon, Kitchener, Lenin or Nazism's sicarians had been devoured down to their entrails by this fire of God before or instead of proceeding to these executions? How can this violence, which the millennia didn't extinguish, be abolished by other means?

A parallel: The abolition of the death penalty

Rightly, a possible conception of peace would consist in defining it as the abolition of a collective death penalty, one suspended over the heads of every combatant, voluntary as well as involuntary, generally the children of those who declare war and who condemn them all to die at an early age. Consequently, the abolition of the death penalty, in the usual sense, plays, in relation to the individual, the same role as peace in relation to collectives. You shall not kill; no individual, no collective, has the right of death over anyone. Here: you shall not kill your sons anymore.

At a date as shamefully recent as the date it consented to the vote for women, my country decided in favour of this abolition in 1981 after political and legal proceedings lasting nine years led by Robert Badinter, blessed be his name. Actual, this resolution is spreading two thousand years after another, virtual, one evoked by the Passion of Jesus Christ, who died, the Christians say, in order to erase the sins of the world, a dogma which in their eyes designates the victim on the Cross as the last person condemned to death for all time: the cessation of the multi-millennial murder of the

Son. His divinity, indeed his eternity, can be measured by the fact that he takes upon himself the sum of the violences of History; the slowness of History can be assessed by the two millennia separating this event from the implementation, still local, of its message; lastly, the new hominescent assessment of life can be divined by the fact that we now think that no one nor anything can take it away before ordinary death. No doubt this is the elementary answer certain contemporaries found to the questions posed by the two bombs, the technological one and the social one. On the other hand, can countries that still enjoy bodies contorted by the electric chair be called civilized?

Let's now note the exact correspondence between the contemporary avatars of death, such as I described them at the beginning of this book (collective or personal, global and atomic, objective and read in cells), and the contemporary avatars of these abolitions, pacifist and judicial, while waiting for life to be defined anew in the laboratories and courts. Is the anthropology of death gradually changing, giving way to another culture, one that's better renascent than Renaissance culture and more different from every previous culture than Greek culture was from Egyptian culture, our culture from that of antiquity or historical culture from prehistorical culture? Those who no longer have the same death no longer live the same life, no longer think in the same way and don't see the same world or the same objects. Who still enjoys, and publicly, the death of his children? How should we define war? As the clear-headed decision made by the belligerents, all of them in full agreement on at least this point, to make the generation coming after them fight one another to the point of extinction. War: the multi-millennial murder of the son.

For your safety ... the end of risk?

The increasing importance of logics of precaution and insurance calculations measures both a trend of increasing intolerance of violent death in Western societies and the new price, almost infinite, placed on life. How can we assess its worth? Lost, from what is said, the sacred returns to the exact place we thought it had left: it was born long ago from human sacrifice; today it forbids even the risk of dying. In order to conquer its mastery of the seas, the England of the nineteenth century accepted two to three shipwrecks per day without shuddering: it paid for the death of its competitors with the death of its children. In order to obtain a power that would expose their lives to assassination attempts, tyrants killed those unknown to them and those close to them. Who today accepts the scandal of dying, even from a deadly disease? Persuaded that they owe us health, we even attack our most effective allies, doctors, surgeons and anaesthetists, before the courts. Life used to be considered, even in my youth, as a risk, a chance, an adventure, whose uncertainties merited incurring the perils of fortune; it has now become owed and a right. So who will deny this progress, even if it comes to us from a recent and burdensome guilt, from an original sin that has reappeared?

The parallel continues: in minimizing risks and building a prophylactic universe through technologies, law, administration and morality, we are demanding a certain abolition of the death penalty for sailors, dam builders, assembly-line workers, travellers and old people. Read the adventure novels of our grandparents, in which ten sailors die from the unwinding of a capstan; the boat and the text continue the voyage, replacing them with ten different hands, without a word of regret or pity; a few decades later, Air France stopped the Concorde, a plane without any catastrophes, at the first disaster. Experts search for black boxes instead of and before accusing the captains, who, as a result, commit suicide less in the event of shipwreck or plane crash than the captains of yesteryear. The responsibility lies in this tool, so new it would require pages of legal and moral philosophy, more than this responsibility now putting the life of the leader at risk, therefore in an object more than in an existence. The death penalty that hung over the heads of responsible parties has once again been abolished.

Lastly, we accept bodies lying amid car debris or planes crashes better than if they fell from bullets: brutal relations with objects replace the deadly relationships between subjects. This progressive objectivization of human relations will soon be revealed to be a variable of change.

Public and political violence

Just as relations between nations couldn't be thought, much less practised, without the ever present horizon of war, in which partners or adversaries at least came to an agreement on the necessity of murdering their sons in mass, so nobody thought or could practice politics without experiencing or balancing the power relations between classes and social violence in its nascent state. Through the eloquence of his displayed body, it was necessary to know how to subdue a group in fusion and arbitrate conflicts of interest when they became exacerbated. Forced to exercise physical courage, a potential victim exposed before the crowd on fire, the political man risked his life; his personal death hung over his head. What we call charisma thus shines with a halo of terror, courage and comedy mixed together. But, once again, the contemporary collective has abolished this death penalty, whether virtual or real. The assassination of Kennedy and the attempt on the life of the Pope date from a return of prehistory that's as intolerable as mafia violence. Reciprocally, who would tolerate the police opening fire on strikers? Abolition of the death penalty for class struggle.

Consequently, domestic politics, as well, has lost the thickness of reality conferred upon it by violence since the foundation of society. The Capitoline Hill used to dominate with all the more height because the neighbouring Tarpeian Rock permanently threatened the life of the potentate; he drew the majesty of his throne from the terrifying dizziness of the cliff the crowd could throw him from. Have we also abolished one of the most archaic death penalties, that of kings and leaders? Yes, for reciprocally, politics flees the places and times in which this social violence becomes reduced. When our representatives, under damning testimony, plead 'responsible, but not guilty', they are measuring exactly the crisis of representation; on the political stage and for the old logic of fighting, politics plays without risk, therefore without any greatness or meriting any respect. If these transparent puppets don't show any courage, what use is granting them power, our old ideology repeats to us secretly? But when we seek our safety at all costs, why shouldn't our rulers imitate us by demanding their own safety? Why shouldn't we have representatives who resemble us? Direct democracy, without any intermediary, with public elections on the internet, will also derive from peace.

Hence the erosion of the state, even more the erosion of politics, formerly 'hard', in favour of three powers which possess the three main components of 'soft' discourse and therefore share an even softer place, without any counter power: the media, via the seduction of words and images; science, via the value of Truth; the judiciary and administration, via the performative function of language.

Soft communication: Rebalancing via the imaginary

Communication discourses, for their part again, measure the various reductions of polemical relations with as much exactitude. Have those who complain about street attacks resulting in two slightly wounded people lost all memory of the world war and the totalitarian politicians resulting in several million corpses? I don't remember people talking about violence so much when it was raging everywhere; to become a problem, it had to die down. The reduction of this 'hard' even conditions, no doubt, the increasingly hard violence of the 'soft'. We must therefore compare the new era of victimization and zero risk to the era of communication. The latter era doesn't so much spread the politically correct, stripped of aggression, as the trend of lower aggression again conditions the spread of messages of all types. So everything happens as though contemporary Western humanity, whether individual or collective, protected itself thrice over: on the one side, through a transfer onto objects, on the other, through the compact walls of signs, through the hard that permits the soft and through the soft that multiplies of itself; lastly through the exportation of deadly behaviours into other countries.

I have long supposed that violence in groups obeys constants similar to energy constants. Just as the exact truths of mechanics and thermodynamics are founded on a stable quantity of force in the Universe, is politics likewise founded on a permanent volume of violence in collectivities? I can't prove this of course, but, even if I prefer to forget this pessimistic thought, I live and think as though I could demonstrate it, in order to try in this way to better understand, prepare and perpetuate peace. Through the conservation of such a constant, one can imagine that the increasingly aggressive representations in the various media – television, novels and the cinema – are trying to reimburse society, anciently athirst, for the blood it is missing, for the mortal risks it no longer runs and for the victories it no longer wins over the corpses mangled on the fields formerly said to be of honour. Thus distant wars, mass massacres, murders, catastrophes and accidents fill the newspapers, whether written, oral or televised, with similar corpses, journals that are repetitive and monotonous since they mistake this bloody oldness of the world for 'news'; indeed, ever since Cain and Abel, what has been new? What moreover is the hero of almost every film playing at? He holds a weapon in his hand, sometimes punishes, always kills. Indignant, for his part, at the spectacular profession, the intellectual writes commentaries about charnel houses; is he, proud, advancing into humanitarian causes, or is he creating his publicity, like the presenter, actor and channel, by means of terror and pity? The media world makes itself glorious from showing murders and wars while claiming it is in this way contributing to reducing them. In fact, this display increases its audience while flattering the most ancient logics, since Aristotle said that terror and pity, precisely, form the best motivations for tragedy. Since the Middle Ages, no one but newscasters and TV station directors have remained disciples of Aristotle; the media remain among the rare places in our society not to have abolished the death penalty yet. Hence their tremendous power and archaic and pagan glory.

The media as more productive than translational

Far from translating, even with some inevitable distortions, reality, whether historical or social, the media create a world that we, virtual animals, tend to believe is more real than the real; this continuous spectacle of violence and death therefore leads us to forget that we have lived in peace precisely since television invaded our homes, which war now surrounds from all quarters. And therefore no one will truly understand this page or my arguments. The West is thus covered with such a veil of melancholy that books and discourses no longer produce anything but these shadows in a time that's exceptional for its duration of peace. The dark virtual produced by these reproducers wins out by far over the evident appearances of everyday life: there is no longer even any ontology anywhere but there. But this blackness occupies the images of the privileged, who live in a white reality. The sons of well-off families enjoy violence more than the children of the poor, who are confronted every day with its sorrow. The impoverished like fine feelings; the wealthy cultivate evil ones, sometimes under the scholarly name of critique.

But who conversely would complain about seeing crimes only in effigy? For, in total, bullfighting, in which a few bulls per year are killed, is preferable to gladiator combats, in which males slaughter one another so others can enjoy it; battles between rare heroes, such as the Horiatii and the Curiatii, are preferable to the mass putting to death of a total war; soccer and rugby, in which players strike a ball, whether round or oval, plus a few blows cracked down on by the referees, are preferable to the blood of formal bullfights; and to finish, preferable as well are symbolic spectacles in which the extras get up after a battle in which a vermillion-coloured liquid flowed in torrents. Who can deny the social progress? Who can deny the general tendency towards the abolition of the death penalty?

The death and the survival of the hero

Have we ever seen three generations follow upon one another without having had any direct experience of war, save war considered in other lands, and living it like a spectacle of distress ethics demanded be stopped? These behaviours are transmitted to children. In times past, during our rearing, we likewise received formulas for survival in the event of extreme violence, and we learned about heroism right from our suckling of the cultural milk, in Homer, Plutarch, Tacitus and Corneille, with Hobbes or Hegel taking over the reins. The hero remained the model, sabre in hand on the fields of battle. Who still admires with all his soul this useless braggart whose boasting spills blood? Should we see, in the end of the hero, an unnoticed reason for the disuse into which literatures have fallen? Let us be delighted at the obsolescence of this swaggerer. The ancient kind is no longer taken up except as cardboard in the commercial series hastily shot in Hollywood. We have abolished the exploits of the blusterer-killer, Achilles or El Cid, and conversely we have glorified the hero-victim, Jew or Christ.

So let's distinguish two courages. For there is still a hero. While the female gives and protects life, certain males, assuredly, but a few women as well, still have a need to grapple with death. Why? I don't know, but I live things in this way. Life can't be taken for value; it only measures it. The value of life is therefore evaluated by the question: for whom, for what would you give it up? So say instead: the value of death. It's not a matter here of killing, an act that only gives a despicable power, the desire for which we share with animals in hunting, but of directly facing dangerous objects: the high seas or high mountains, great colds and deserts, but also solitude, exclusion, independence of life, freedom of thought, all of them objective obstacles suited for testing the limits of a body common to all the humans that we are. Without this specific courage, there can be no human challenge. Therefore I am reversing what naturalists know and say: animals kill one another and thus perpetuate evolution. He who no longer kills thus promotes a different time, a human one. Might the ending of the hero then take a noble place in the process of hominescence?

So there are two boldnesses in the face of the reaper, that of baboons, of ibex, of Captain Fracasse and of Tranche-Montagne, whose boastings dazzle the audience; the family of mythomaniacs doesn't die out so easily.[2] When the adversity of a hard life hasn't already taken it upon itself to teach it to you, the second boldness occurs when faced with bad weather on a glacier, at the foot of a vertical face or in crossing a wide and turbulent ocean. Thus the first traveller, quitting his home to cross a forest massif, high crests or the sea, launched humanity into its own history more surely than the killer, who only reproduces bestial hunting; the first one confronted objects of the world. He no doubt knew for what and for whom. But, in the course of a battle that takes you by surprise, the other courage always and again returns; replacing the one that kills, it consists in interceding to make the killing stop and also in taking it upon yourself never to take vengeance. Others are never killed by anything except fear. Matamore, the braggart, gives himself over to grand discourses and broad gestures because an underwear-soiling fear seizes him in the guts. The hero confronts situations and objects peacefully. And solitude as well.

An objective cause: The object

I therefore believe that the main cause of the advent and the length of the peace resides in the hominescent status we grant to the object. Isn't the motor of this long peace present in the proliferation of all those objects consumption causes to flow around us like a flood, as if we were transferring towards this new objective universe the subjective violence that in the past and recently we only knew how to transfer on to our neighbour? I lean towards this solution. Before the fall of the Iron Curtain, I saw, in America, Soviet travellers faint upon entering the department stores: their bodies couldn't endure this total slackening of desire.

The ancient bloody catechism and the behaviours it induces in fact presuppose or produce a complete philosophy of the relations between subjects, on which the relation of war depends: objects don't exist independently but only insofar as certain subjects possess them. From a very early age, we only desire or want some thing because of our relations to the people who hold it. Language says: *choses* [things] come from *causes* [cases or causes]; the first of these two words comes from the second one. So first there are causes to defend, causes of disagreement, cases to plead before courts, and things depend on a decision acquired after a debate, a battle, a war or a trial. Objectivity ensues from a dispute to be settled.

These relations of comparison or jealousy measure desire and value. In the first paradise, the tempter didn't describe the apple as being delicious, since our Edenic parents were up to their ears in delicious dishes, but made it desirable by suggesting to them that, if they ate it, they would become as God. It's not about fruit but about measuring oneself against the Other, against his greatness and power; it's not about the thing but about relation. All the misfortune of the world, all its violence comes from comparison. Thus the object is mixed with so much subjective that wealth must be understood to be this subject–object, this quasi-object: the stake of power, the fetish of adoration, desirable merchandise. All things are mixtures of the three. So there is no objective existence yet, only intersubjective relations. Yet our safety protects people more and more, and we produce more and more things. We are reversing the most ancient relations between these two words.

Political animals

Observe for example, when hunting or gathering comes to an end, these tigers or chimpanzees eating. None of them grabs a single morsel before the dominant animal does. After his meal, this one and this one alone gets to decide who is going to eat after him and which after the other. This sequence of power and violence forms their society. The social is linked together and hierarchized via these series. The prey or the banana, which play the role here of objects passing from the one to the other, only exist through the grace or power of the hierarch. There is no treasure except the one designated by the leader to be such and which he allows the other to acquire and consume. Desire increases with the wait in the sequence and produces in it every possible squabble, quickly crushed by the leader's power. This latter therefore lights the fires of desire so as to better put them out. The supreme captain of the armies before Troy, Agamemnon was likewise in charge of the division of the spoils plundered from the enemy between the envious Greek petty kings, whose rivalry lit the domestic war preceding the foreign war against the Trojans, which itself is explained by the rivalry of two leaders in love with the same Helen, whore-spoils, the cause of the war, a beautiful thing [*chose*], subject–object.

Enclosed therefore in the pure intersubjective, in the exclusive political, animals and their immediate human successors have no objects except for those included in the power of a single creature and enriched by reciprocal hatred. The delights of the prey, Helen's beauty and the gold of the spoils shine only from the fires of this desire. Understand by politics or power relations the management of this violence and the monopoly of its laws. So those who share violence or appropriate it for themselves alone define the very existence and value of objects. Relational strategies precede objective behaviour. In hominian evolution, the knowledge and practices of the intersubjective and its strategies precede all objective knowledge and practice: the social sciences precede the hard sciences. The more politics there is, the fewer objects there are. Conversely, when the objective increases, politics decreases. The more rivalry there is, the fewer objects there are. Does war decrease when objective existence occurs? Yes.

Man, the objective animal

No, we are not political animals; yes, animals, for their part, remain political, and we, having long remained among their retinue, are becoming, little by little, objective animals: the object as such creates that which, in hominescence, quits animality. Enclosed in the intersubjective, in the prisons of coevolution, animals have no objects, merely a *Lebenswelt*, a vital world of niches and prey; hence their enclosure in the time of struggle, whose deadly cruelty Darwin himself bemoaned. Who was the first to delimit a niche in order to say: This is mine? Every living thing, and millions of years ago. The continuation of this gesture has been carried on archaically through its defence, coevolution. Consequently, we were long and often still remain political animals, Agamemnons, Ajaxes, Achilleses, despotic petty leaders surrounded by warriors who were so close to animal lineages, drooling with the desire that made them draw their sabres. We didn't leave off being tigers so easily. That began well before the Trojan War, probably from our own emergence in the African East, even though we didn't know anything about it, and lasted up until the Hiroshima explosion, even though we couldn't yet hope that this last quasi-object (last because filling the entire horizon) had finally sounded, as a world-object, the definitive end of animal struggles: a fine social tool.

The final loops and a visit to the sewage fields

I am attempting to describe the process in the course of which we came to believe, little by little, that objects exist independently of subjects. For tens of thousands of years, at least, we have been carried along by a self-perpetuating intersubjective loop, persuading us, with its spiral motion, that goods had to be divided up since the set of objects reduced to the sum of what the set of subjects had in its possession. So waging war was necessary in order to obtain a share of this finite lot, a lot defined by the number of its possessors and the more desirable the more this number climbs and the battle burns. Does an inverse objective loop carry us along today, self-perpetuating, as I hope, persuading us first that objective goods exist independently of us in an infinite amount and next that we can produce, at leisure, an equal quantity of them? Consequently, abundance would depend less on peace than peace would ensue from superabundant objects. Would it result from the end of their rarity?

One must have lived in deprivation, austerely, before the beginning of this sudden flood of objective consumption or have dwelt a long time in poor countries or in deserts, for example, to shift the conception we have of wealth a little. No, it doesn't merely have to do with property value, financial value, stock value, bank value, jewellery value, and therefore with the relative scale of possession, but precisely with the raw quantity of available objects, even if we don't possess them. Over the course of the last half century, the maximal gradient of change concerned precisely this number, produced by industry, technology and science. Writers and poets, Ponge and Perec, suddenly began to talk about objects as such.

But we went from their rarity to an excessive inundation, proof of this being our worries regarding garbage. We are overflowing with so many things that we are risking, tomorrow, drowning in a sea of trash cans and a flotilla of hopper barges. The sewage fields over dozens of hectares on the outskirts of cities would deserve the name of suburbs or places of banishment [*banlieues ou lieux de banissement*] better than the other ones. In the most logical and most metaphysical of all his dialogues, the

Parmenides, Plato asks if one can form the Idea of excrement and offers the examples of hair, mud and dirt – the waste of the land and the body. We wash ourselves enough now in order not to even know what they look like, but we substitute our debris and rubbish for these examples. Cleaned up, the new body, the new agriculture, without much mud, the new life, forgetful of death, the communications without obstacles, transferred what our philosophies can't conceive any idea of into landfills and garbage dumps. Across their fumaroles and stink, armies of aggressive rats show, in these landscapes from which philosophers absent themselves, damage and harm, the inverse image of our new peace.

When objects become substituted for subjects, sewage fields advantageously replace the fields of honour; do you prefer rows of low white crosses, after the battle, in lines as far as the eye can see, as though still at attention, to these chaotic plains of night soil? Those who still like to fight resemble the giant and cruel rats fattened with filth who reign there over what seems highly desirable to them. Our old history remains in these places, where the mushroom cloud veils the sooty sparks running over the mire, where all our world-objects, gathered around the horizon's perimeter, light up and overlook these myriads of sewage.

The end of the perpetual?

Before we downgrade them, said objects support us in our existence, transport us, treat us, improve our comfort, soothe our pain, calculate for us, remember in our stead, communicate without us, occupy and build our houses; even the increase in life expectancy, to my lights, passes for an objective, relative and progressive abolition of the death penalty. How would you expect these objects not to end up defending us?

Lastly, since the atomic bomb culminates as the last quasi-object to occupy the entire horizon of human desire – the sum of hopper barges, the ultimate trash can, the universal sewage field – will we also wage war with objects, without humans? Of course. The after-sales services of arms dealers (when will they appear before the international courts for crimes against humanity?), perfectly organized for commerce and advertising, will dump their hellish products onto the battlefields, after having adjusted their delicate mechanisms, in order to make them go off from afar in the absence of human armies. The so-called professional army will have joined in retreat the conscripts recently condemned as the gladiators were long ago. The abolition of the death penalty will have touched even its executioners.

When will this ever so long peace end? With war, assuredly. If today businesses help themselves to laboratories, if politics intervenes in the choice of research and their financing, if lastly the scientists themselves, hungering for glory, sing in the media what suits their renown, interest or pressure group, then our society, thirsting to put subjective relations back inside the most objective relations, will no longer find itself protected by the object, that guardian of peace. The forces stemming from evolutionary times are always on the watch in us in order to bring back the bloody eras of dominance via combat. When these powers succeed, tomorrow morning no doubt, then world war will return, which, as usual, will cost infinitely more than the price the rich could have paid to avoid it. This war will start again under the impetus of these interests, an impetus suddenly combined with the expectations of those who will become their adversaries, the billions of disadvantaged humans who will not be allowed, by our injustices, to share with us the hope, as close as it is disappointed, for this possible perpetual peace, a now tangible stage of the process of hominescence.

Lost and re-found projects

That our societies no longer have any projects, that they seem to believe in concepts as empty as the end of History, a postmodern era or the perpetual present, that, benumbed in memories and nostalgia, they no longer conceive of anything but celebrations in museums, that the pens of the learned, who are losing social power due to multiple changes, brood ink, what could be more normal since every project they have ever conceived and carried out, formerly and recently, started from war, pillaging and extermination, from invasion and imperialism, since every one of their thoughts has only ever given itself as its avowed or secret goal the advent of *Homo terminator* or of the dominant one? Is there a harder drug than spilt blood? Dulled, stupefied by the sudden stoppage of a time relying on its narcotic effusion, our societies seem to tolerate poorly this abolition of the death penalty, an abolition that's so general it touches the soldier, the captain of a vessel, the victor and the victim, the traveller traversing the ocean and the passer-by crossing the intersection, the mason equipped with his helmet, the miner and the alpinist, the king and the politician, the striker and class struggle, in brief, Mars the warrior, Jupiter the leader and Quirinus the worker, in sum everybody and life itself. What absence of pity impels them to accuse themselves of decline before these infinitely valuable new lives and new bodies, this universal pardon, this putting to death of death, this peace? With the advent of this general abolition, how can we tolerate the universal fact that the majority of humans do not have this right?

Envoi

A nice work in progress for philosophy: thinking a time without dialectic, constructing a history founded on life, creating values purged of resentment, generalizing peace, which never comes about from violent protests against violence but through converting aggression into energy mobilized towards a project. Of course, peace can't be taken for a goal, or rather it is only considered an end in times of war. Once there is peace, what should we do? Peace, in itself, reduces to a means. For what project? First for sharing it, for only a minority of humans enjoy it today. This work will still take centuries no doubt.

If violence easily spreads so quickly and so far that this expansion can be taken to be its very definition, and if its representations and even the heated protests against it objectively serve its dissemination, spreading peace consists in going back up the course of entropy, and therefore the time of its fall. This reversal defines the work and the unexpected invention, and therefore the thought, in general, that go back up the mortal chaos and the flood of blood. Project peace next onto the inert world and onto the living things with which we maintain a universal war. Therefore draw up a symbiotic *Natural Contract*. The philosopher's duty consists, during these beginnings, in conceiving this work and this thought, this new recountable narrative, in constructing another history, in going back up the new time.

Dating

During this month a global revolution took place, or at least a Western one, which didn't attack anyone nor did it recognize any adversary, whose stakes neither Marxism nor Leftism understood, an earthquake that believed itself to be political and whose political analysis didn't understand or explain any gesture or any action. Besides, there wasn't any action. So what happened in May 1968? Assuredly nothing if not the act of birth of this new world without any relation to any other since the beginnings of humans and societies. For every index shows that this date, 1970, to round off numbers, slices the century and perhaps our History; the event escaped it.

Mars, Quirinus, again

Nuclear bombs, in the middle of the last century, imposed a deterrence whose effect not only warded off the strike of some designated enemy, but conflict in general as a means of resolving crises or tensions. For a world conflict, maximal, would cost the elimination of our species. At three-quarters of a century, did a certain warrior past leave us?

We were at the epicentre of another major earthquake, one that transformed the relations of the West to the land, to production, to exchanges, to wealth, to consumption and to comfort while the shadow of Karl Marx, projected by the furnaces of a vanishing industry, was disappearing in the midst of the surface convulsions. This book dwells on the least known part of this second quake, in my opinion the most profound part. Stemming from the Movement of Catholic and Agricultural Youth, a new generation of farmers, my generation – for the moment, I'm only talking about France – came into conflict with the old generation because of the new problems posed to cultivation practices by mechanics, agronomy, biology and biochemistry. The rural exodus deported the major part of families to the city and in a few decades overturned the ancient percentages of our populations and our archaic relationship to the land. In the middle of the century, the last shred of the most ancient past, in its turn, left us. Note that the term 'green revolution' also dates from 1968 and that, to leave France, M.S. Swaminathan, a geneticist of that generation, said that his country, India, had been self-sufficient in food since 1975; the times agree, roughly, for the entire world.

The collectivity the most enrooted in hereditary archaism overthrew its internal relations. While, ever since the Neolithic no doubt, the old man on the farm had commanded silent young people, in the 1950s, we find the son making his voice heard by the father, who didn't see the new world coming the way, just now, the general no longer heard the new lieutenant, become an atomic engineer, or again the way the deflationist financier of the French Third Republic went into a panic in the face of the sudden growth, inflation and consumerism, or the cardinal in the face of what the curate of the neighbourhoods was claiming. Suddenly, the young, through rigorous reason, were persuading the old that their experience no longer had any value. Once again, the son doesn't want to die anymore.

Historical, events speed up: from 1959, meetings and violence in the street began in Brittany; on 8 June 1960, thousands of farmers invaded the prefecture of Morlaix; the movement extended to everywhere in France. How is it possible that we remember the revolt of the students of May 1968 and its unrest without any wounded with so much sharpness while forgetting the revolt of the livestock farmers, which caused, in 1967, hundreds of wounded in Redon, Quimper, Rodez and the Basses-Pyrénées? Proof that in those days we were already in the process of losing the land, which we all formerly participated in, whereas what was happening was announcing the future.

Jupiter, in addition

So three earthquakes: the first one concerned the relations of violence between humans and its multi-millennial solution via war; the second one concerned economics, production and agrarian exchanges; the third one concerned religion, culture and education, from the Catholic Second Vatican Council to 1968's university unrest. Is this a question of one and the same phenomenon? Yes. Here it is. By themselves, the physical sciences made our destructive powers grow to the limits; agricultural production increased to the point of emptying the Western countryside of humans, principally due to the agronomical, biological and chemical sciences. On balance, the overturnings that happened in the century were moved by exact knowledge; but this knowledge only determined them by passing through the evolution of the body and of the world, deeper.

Neither the farmers nor the students live anymore in the same culture – let's retain the same word for both senses, as long as we still remember them – as their predecessors. Nor does anyone from now on. While awaiting the biotechnologies – agrochemistry and agrobiology opposed their green revolutions to the traditions of cultivation practices [*façons culturales*] in the same way that multiple scientific revolutions clashed with cultural memory. Just as for the planet Earth, there exists, for history and social formations, a tectonics governing this collective time, a time as complex, dark and silent as the time of things. This is the subject of this book: What are the plates that, beneath our feet, roll the deepest and the slowest and which historical ruptures, superficial, reveal as volcanoes?

Should a century see all its conflicts flare high and universally to the point that they reach the maximum of horror for a minimum of utility, to the point of even annulling their effects, under the influence today of atomic weapons, and should, after such a vain explosion of violence, the holders of the power of absolute destruction only envision war to be a possible solution to crises through short-sightedness; should the same century see the one universal Church, and not only by name, undertake, from 1962 to 1965, its most important updating since its two-thousand-year-old foundation, and, not long after, Western agricultures enter, from 1966 on, into a crisis in which they again lose their numbers; then we must remember, at the limits of written history, that the farmer, the priest and the soldier have continuously held our societies since the Neolithic, and therefore conclude that our relations to the land, to violence and to the sacred reach the lowest and slowest plates in time; should these plates move, then historical times will renew their décor. Should, in addition, the new technologies transform the social bond, then, truly, we will find ourselves in the presence of the most violent earthquake in more than five thousand years. The unrest of 1968 ensued as echoing repercussions of these three major shocks, more powerful than this unrest, but of the same family. It has less to do with history than with anthropology and less to do with this latter perhaps than with evolution.

The twentieth century marks the end of a certain reign of Mars, the traditional warrior, having reached his maximal power and therefore pulled towards new solutions, of Quirinus, the ancestral agricultural producer, for whom, now, abundance exists next to destitution and comes after rarity, of Jupiter, the priest stable in time; the trilogy of our forefathers and of the oldest gods is vanishing. We don't produce or work the way we once did; we no longer fight like our ancestors; we no longer know or pray with the same certainties or the same hopes as they did. Changes that are so decisive they uncover the origin, an origin the new events whose burns we have lived through have received from the most buried of archaisms, the way volcanoes put the earth's surface and its red and black entrails into short-circuit. So under what has been called the substructure, no longer fixed but in motion, lies this tectonics in which the plates of rites, brawls and fruit slide over each other and break.

A strange world, young and old, in which our contemporaries, fearful in the face of the new, get lost amid the archaisms that return and dominate so often. Such a dangerous rebirth fills one with enthusiasm and makes one positive and optimistic just as I am and was, a witness, an actor, in the middle of overturnings whose unity we can now perceive: the end of agriculture, in the oldest sense, war and death camps, the rise in power of economies of communication, the crisis of knowledge and education.

What happened in 1968?

We experienced the powerful feeling that the old world was collapsing without anyone pushing it into the nothingness; we had no opponents or enemies, and no one contested, except for the elite and the majority, blind as usual, the arrival of another world. The May event vanished as quickly as it came, as though it had never existed; the daily grind seemed to resume as before. But it soon became evident that precisely nothing was continuing on as before. The sons had reason for being right over their fathers. The new reason has already been dated.

I have known this reason since the 1950s. At the start of the 1960s, I had raised Hermes, the god of communication, to philosophical rank; I sensed Prometheus' fall, of course, but especially the end of the Neolithic. In order to express my thought better, I shall go back to an even earlier decade, the one preceding the Second World War, from which I get my first sure memories. I can testify to the fact that no gesture, whether one of sowing or benediction, no act, whether medical or political, no profession, whether smith, wholesale or retail grocer, saddler, farrier or empirical veterinarian, no social, sexual or pedagogical idea now survives of that era and that conversely I can't put any of today's gestures into this more than old framework. I don't know of any means to translate one world into another when they don't have in common a single landscape, whether urban or rural, few fabrics, not a shoe, few foods, probably not even a meaning. We who cursed, in sung canon, the bell-ringer and the regular soundings of his bell, wouldn't have endured for three minutes the din of engines and radios. Landscape gardeners because peasants [*paysagistes parce que paysans*], would we have endured the recent hideousness of our city entrances?

But, hobbling along, this quasi-history persisted up until the 1960s, precisely. It only collapsed at the end of that decade, when the polyneism reached the threshold of percolation, in the sense I have given these words. The death throes of the old world, whose beginnings no doubt go back to the moment when humanity invented agriculture or even to the moment when the human body such as it was worn by the peasants of my country was formed, therefore occupied the three decades from 1938 to 1968, that brief interval separating the Spanish Civil War from the aforementioned student revolution, if you want to talk like historians. The explosion of two atomic bombs, the eradication, in the full sense, of farmers, the appearance of a new human body and of its relations to a now global planet cadenced this interval as a result of technologies and objective sciences, physics, chemistry and biology.

The loss of politics?

None of these stages concerned politics, still less economics; all of them stemmed from inventions issuing from the hard sciences, atomic engineering, information theory, pharmacy and biology, affecting physical or vital variables, themselves plunging into evolution. The year 1968 showed the body in glory and travelled the entire world, that's all; everyone sang and laughed, that's all. Only a few backward groups, club in hand, imitated the fascists and Stalinists. Trained in economics and politics, seeking debates and adversaries, the historian understands poorly this gushing forth without any dialectics or class struggle, without any elective stakes or representatives, without any real opposition, this revolutionary movement without any revolution. Demonstrating in the name of exotic or old political ideals, without any relationship to its real conditions, the actors of the movement didn't understand much better; without knowing to whom or to what they were referring, they believed themselves to be in contemporary China or in the czarist Russia of the beginning of the century. The event greatly exceeded the interpretation of its performers, stagers and chroniclers. Apparently with good reason, we ended up scorning this non-event, qualified at the time as a spectacle by those who were less naive.

Yet how can we minimize the importance of a date that reached the universe, testifying for the first time to the emergence of global actions and opinions since not many capitals at that time dispensed with reacting? What then is a universal non-event? No doubt a circumstance whose datum, whose conditions and meaning no one reads. In fact, this circumstance exceeded History such as we conceive it. It referred to dates preceding agriculture, immemorial, writing, archaic, medicine, antique, politics, ancient, and economics, so late in coming. It translated the liberation from bodily constraints, from the spatial conditions of communication, from the local frameworks of technological action, from the old dependences with regard to things which never used to depend on us and which, now, depend on us: the subjective body, the objective world, and the others or the collective. The festive atmosphere of liberation struck the hour of the new liberty, which wasn't acquired by overthrowing a prior personal, collective, in short, human power, but by certifying that past alienations, both physical and vital ones, had vanished. At the very moment when the students were shouting 'everything is political,' nothing was so any longer.

Consequently, Jupiter, Mars and Quirinus still mask from us, through their Indo-European narrowness, the true span of the event that certified the death of these three gods. Global, physical and biological, this event exceeds politics, economics and History. Affecting the relationship of hominity to oneself, to others and to the world, it concerns the process of hominescence. Blindly, May 1968 dates it punctually.

Generalization

What, invisible and innocent, this terminal point has made visible, the entire century and its long trail of abominations has shown even more and in a wider theatre. Europe's cynical suicide in 1914; the globalization of war; the eradication of peasants in the Western countries, a doing away with that began with their useless sacrifice during the first world conflict; the devastation of colonial countries; the twin totalitarianisms of the Soviet Union and of fascism in its three versions, Francoist, Mussolinian and Nazi, which dominated Europe and the century from 1917 to 1989; the Final Solution and the extermination camps; the atomic bombs of Hiroshima and Nagasaki; the birth and putting to death of the third and fourth worlds; the gigantic farce of the subcultures issuing from totalitarian ideologies first and from the capitalism of the United States next, this latter all the more saturated with dollars for starving humanity while watering it with gobs of ugliness. This hideous sequence, which made the twentieth century one of the worst in History, at least due to its effectiveness, appears to me today to be the set of turbulences that resisted as disorder this irresistible process of hominescence, overturning knowledge and the effectiveness of practices, as well as bodies, the planet and human relations.

Since the reaction allows us to evaluate the action, these regressions measure the crossing of a tremendous threshold. Has the parallel between these ruptures, due in part to particles and the decoding of the genome, and the condemnations of physics by Lenin and of genetics under Stalin ever been noticed? Were they afraid of true causes? Do we still observe the reduction of science to politics under totalitarian regimes, Nazi and Stalinist, as well as in the correct thought of today? So when our relation to ourselves, to pain, to food, to reproduction, to the land changed, when our ties to others, and therefore persons, groups, societies at the same time as the world transformed, these actions brought about backlashes that were all the more reactive because no one could resist an impetus no one understood. Didn't the progress of the sciences during the Renaissance and the redefinition of humankind that ensued produce the Spanish Reconquista and the crimes committed in the newly discovered America at the same time as the Wars of Religion in France and the Peasants' War in Germany? In a river, eddies form that are all the more violent and spiralling as the flood rises irresistibly; the slow advance, under the land, of tectonic plates explains eruptions and earthquakes. Thus, and this book often refers to this, local innovations in the sciences and technologies, suddenly having repercussions on the whole of society, unleash tremendous returns to ancient ages. The final loop: thus, and indirectly, the process of hominescence unleashed these wars and violences and, conversely or in their turbulence, brought about our new peace.

Prescription

In this century which I hope has vanished, certain decision-makers I have a memory of, their collaborators and their sycophants, probably hadn't entered into any more of an alliance with evil, death and infamy than the decision-makers of today or the kinglets of times past, but a vast flood, which neither politicians nor philosophers had known of or been aware of, which besides we didn't yet have a clear knowledge of, laminated us and them. Neither more nor less human than the ancients or our contemporaries, they became, through their crimes, the actors and victims of a tsunami that was so far above the ordinary water flows of History that it reached that alert level I mark by means of hominescence. From not having been able to change hominity nor manage with prudence and wisdom a flow they hadn't mastered or understood, they committed the worst of crimes.

In assessing, on the evening of my existence, the objective height of this flood, I hope that our generation pardons them, even if they did destroy our childhood, pillage our youth, often make the solitary choices of our adulthood painful and therefore left a wound that will no longer heal on our lives and freedom of thought. May prescription, the foundation of human justice and law, at least descend upon them.

Sciences and cultures

Let's suppose that in Heaven or the Underworld, depending on their merits, the shade of Hippocrates, the Greek doctor, the first one in the West worthy of the name, and the shade of Aristotle, the philosopher of life, meet and discuss illnesses and health with the shades of Galen, their Latin equivalent, and of Ambroise Paré, Laennec and Pasteur, our historical models in clinical medicine and therapies; for good measure, let's add Bordeu, who Diderot relates treated d'Alembert, and Semmelweis to this group. Their different cultures, the centuries that separate them, the experiments and discoveries that distinguish each of their knowledges and practices won't prevent them, enjoying the same view of living things, from understanding each other in order to debate the same concern revolving around the same object. Since all of them maintained more or less close relations with the philosophy of their day, we instinctively call them humanists, experts in strange animals endowed with language and traversed with suffering.

At this juncture, let's suppose a biochemist, a Nobel Prize laureate from after the 1970s; happening to have passed away, this scientific contemporary desires to join this conversation from beyond the grave. Catastrophe: it breaks off. Neither Darwin, an additional guest, nor Pasteur, the first master of microbes, nor even the biologists of the beginning of the twentieth century will in fact understand a thing about his allusions to DNA polymerase, kinesin or nuclear magnetic resonance imaging; conversely, let's wager that he will also understand as little of Galen's logical quibbling or Bordeu's dreams. As an iceberg does from its glacier, the knowledge of the last one in date suddenly breaks away from the common and continuous bloc constituted before him, nay, from the ancient global comprehension concerning patients and humans as well as from the causal relations with suffering. He cuts all ties with the History that binds his predecessors. We neither have nor know the same body.

I can suppose and then stage the same dialogue of the dead bringing together astronomers or specialists of the Earth sciences with the same result. Ptolemy, the last Greek geometer to draw up a model of the sky, Copernicus, who, against him, put the Sun at the centre of its planets, Galileo, whose telescope saw Jupiter's satellites, Newton, the author of the law of universal gravitation, Laplace, who gave us confidence in the stability of the solar system, and Poincaré, who shook this confidence, all of them see and think the same world, even though, from the epicycles of the first to the nascent chaos theory in the last, all of science has changed during the two millennia separating them. Humanists, close to the great philosophical debates of their times and sometimes even at their source, they will understand one another because they agree, without saying so, on the same object: motions in a system. Even if they might completely reverse their neighbour's view, they nevertheless share the same view. As the learned say, they changed paradigms, of course, but within the same world. However, they will not be able to understand anything the astrophysicist, arriving there, might say, preoccupied with black holes and other gamma-ray bursts, designating the great attractor swarming with galaxies millions of light-years away in the depths of the Universe as his horizon, lastly perceiving the light emanating from the two seconds after the big bang. A crevasse separates them. We have changed skies.

But, to my knowledge, Eratosthenes, who was the first to measure the circumference of the Earth by measuring the shadows of sundials placed in Greece and southern Egypt on the same solstice day, would likewise understand, and perfectly at that, Mercator, whose maps gave the Renaissance navigators the best assurances against shipwreck on the shore, Leibniz, a manager of the Harz mines, who pulled the first fossils from these depths, Buffon, who calculated the age of the Earth on forge-heated iron balls, Lyell or Jules Verne, whose imaginary hero journeys to the centre of the Earth ... But the cosmonaut who saw and photographed the blue planet for the enraptured astonishment of all of humanity, but the geophysicist who observes from his bathyscaphe a submarine eruption at the depth of five thousand metres, but the plate theorist who explains volcanoes and earthquakes by the sliding of thick and slow masses ... would only understand the previous ones across the mists of a prehistory and speak to them a language they in return would not understand. We barely live on the same Earth.

Let's now listen, in the same Underworld, to the conversation of writers and philosophers, politicians, shopkeepers, industrialists, who invite, once again, doctors, astronomers and geologists, along with lawyers or professors, all of them in their flourishing age around 1900. Let more people arise among them, such as a Greek farmer from before the fifth century BC, come out of, for example, Hesiod's *Works and Days*, or some Latin agronomist, Columella or Varro, inspiring the *Georgics* or inspired by it, a serf from the Middle Ages, brought to life from the *Très Riches Heures du Duc de Berry*, as though illuminated, a farmer from the seventeenth century or the Romantic age, come from a painting by Le Nain, from a theory by Say, Turgo or Adam Smith, or from the stories of Marivaux, Goethe, George Sand or Zola ... To my knowledge, all these modern decision-makers will be able to debate with these peasants stemming from every age about ploughing, manure, autumn sowing, the cultivation practices suited to vineyards, the way to milk cows and the difficulties in feeding bees or birthing mares; they will manifest the same worries about the vagaries of the weather, excessive rains, long dry spells, summer hails or wintery weather under the April moon, for they have kept the same nature, the same relationship to working the soil since the Neolithic, during which, a miracle, animal and plant species entered into domestication in their own houses. Through this first meaning of culture, we come even closer to humanism, for even if a livestock farmer from the Sahel or an Asian rice grower joined in the conversation, who can believe they would truly find themselves to be not at home? Therefore, a similar group of people in charge today, come out of the schools which know, govern and administrate, must in fact wait at least until they descend into the Underworld to meet some farmer: they no longer have any idea what his practices are. Vegetables, fruits, meat and milk, available at the supermarket, amount, in their eyes, to products for sale or consumption, marginal elements of economics, separate papers in several files nowhere to be found or turning up with difficulty. Having chased away every living thing from our residences, they only live in symbiosis with their fellows or their photos on screens.

But what are all these important gentlemen, even dead, doing together? Doctors or astronomers, kings or rich men, given over or not given over to ploughing and pasturage, all of them have an understanding, without saying so, to exclude women from their conversation, our equals at least, as well as the poorest, their masters in social experience. Neither heaven nor hell prevails against the prejudices of these predominants. Whether it's a question of the sciences or the arts, of creating or deciding, of observing the heavens, of conceiving or of working the land, of leaning over bodies with tenderness, the great names didn't state themselves in the feminine. Humanism, male chauvinism. These illustrious males never recount anything but a half of history, which never recites anything but a demi-humanity and – will we ever know this? – anything but truncated sciences and practices. Just as, right-handed, they use a hemiplegic body and functions split down the middle, so they cut off the female part, no doubt the more precious one, from their experiences and thoughts. In these frightened dialogues convoking masters and disciples, the word 'mistress' still causes a few imbeciles to laugh in the academies. So what faces would these monomaniacs pull when this mistress, who in the absence of annals I shall call Lucianne, appears, more educated, more astute, nobler, less crude, hungering less for power than they do? Those who make her keep silent are now reaching the weighty age of dinosaurs, afflicted with their light brain. Without waiting for the Underworld, women are finally talking in assemblies. But we don't yet hear the voices of the poorest, still likened to wolves.

What then happens when the same sky no longer lights us or protects our rest, when we no longer inhabit the same body, when another sex, finally recognized, delights and teaches us, when neither the same Earth nor the same earth rolls beneath our feet, in sum, when the child culture loses the world belonging to its fathers? A new humanity happens.

I hardly dare to repeat the same scene with the conversation bringing together those who sacrificed to Venus in Greek or Latin antiquity and the disciples of Baal on the shores of Carthage, the sectators of Mithra amid declining Rome and the initiates of Eleusis, the worshipers of Yahweh from Abrahamic times and the first Christians sacrificed to the wildcats in the Coliseum, a few Benedictines or Franciscans from the Middle Ages, Reformation theologians, those messieurs of Port-Royal, in short, Saint Francis of Assisi, Calvin and Arnaud, Voltaire the deist and Lamennais the author of *Words of a Believer*, Léon Bloy, Péguy or Bernanos … What student of letters or of philosophy, having died this morning due to an unfortunate car accident and trained by daily reading of newspapers to only think about persecuted religions in the despised form of fundamentalism and Inquisition, could join in the debates of the preceding people, could give, upon his turn to speak, a pertinent opinion on their hopes, or could, at least, conceive their aspirations towards a spirituality? However far we went back in time, all the way to the Lascaux caves, however far we explored space, everywhere and always, *Homo sapiens* showed itself to be ritualistic and pious, a sacrificer and religious; it perforated the profane with the sacred, prayed and recited litanies, knew how to integrate, lastly, the whole of its acts of love in mystic ecstasy. During prehistory and History, in every climate, the gods have walked, absent, in the shadow of *Homo pius*. Without looking closely at it, one generation has taken away from the following one the contents of this inheritance and deprived it of this accompaniment. Has the spirit of spirituality had its day?

But who among my contemporaries now has read, in his language, or even ever heard Lucian spoken of, the immortal author of the *Dialogues of the Dead*? The theatre whose set I just populated with different shades doesn't give rise to any echo, and no one has grasped why I called the first woman invited Lucianne, or why I evoked wolves in our murder of the poor. A Great Whole dies again, the one we united under the name of culture. The Greek dead man, the Latin dead man, those dead of the *Dialogues* have just died one last time. Their survival was due to oral languages then written languages, to books, to their printed delivery, in short, to memories traced on various recording media, come into being before Aristophanes and the Presocratics for writing, before Erasmus and Rabelais for printing. All of these people I just gathered in a paradisiacal or punitive place, these scientists and these religious figures, these male organizers and wealthy farmers read and knew paper, wax or papyrus, letters, stylus or pen, the upper and lower cases of print shop foremen. I work on a console and hold a mouse in my hand; the text I write or read unfolds on a screen; its processing erases mistakes and struck-through text; my memory runs on the internet and lies in data banks; as many calculators and software do a thousand operations in my stead – I therefore no longer dispose of the same 'faculties' as all these infernal chatterers. From having quit the old recording media, I no longer think in the same categories and no longer have the same memories. How can we be scandalized that the *Dialogues of the Dead* are dead when we no longer even preserve its writings and words in the same way?

Let's close the Underworld for inventory.

Yesterday and tomorrow, humanism

What we have long praised by the name of humanism and which we nevertheless have rarely defined was therefore due to everything those very people we no longer understand understood: to a compact and reasonable world, farmed along the same furrows, defined by certain changing sciences, spoken of in certain languages shaped by time, saved by religions, to a human fate meditated on by bodies in pain that no medicine or remedy could relieve, much less cure, bodies bent with melancholy in the face of violence and often enduring it, lamenting plundering and the desire for power, but above all, but before all, desirous of, hungering, thirsting for beauty. This humanism reverberates in the ears like a subtle music. Who today would sacrifice his life to this beauty we pretend not to be able to define universally? Who would believe that it saves the universe, precisely, because everyone born into this world recognizes the works beauty appears in? Who consequently spends his days and dawns with the harmony of language, with the constellated painting of vowels, with the complex rhythms and tempos of sentences, with the exquisite linking of meaning and form, with the mosaic of form and the music of meaning? This humanism has had its day; I have witnessed, in the course of my short life, its brief death pangs. I intend for us to talk about it, but as in a *Dialogue of the Dead*, we will no longer resurrect it for the simple reason that the world that fed it, that conditioned it, that also resulted from it, has vanished without any hope of return. Different bodies differently sexed move on a different earth and under different skies, composing in different keys.

As a result, we are living through a Renaissance next to which the one the humanists called by this name amounts to a gentle ripple in the flow of time. Under a sky of such an immensity, on an Earth we have just acquired a moving global knowledge of, inhabiting a body not brought closer to the bodies of our parents by any pain, using recording media unknown to our predecessors, awaiting, perhaps, an undecipherable god, what are we waiting for to invent, not a second humanism, but humanism as such since for the first time in the million-year-long process of hominization we have the scientific, technological and cognitive means, via easy studies, effortless journeys, multiple and unexpected encounters and proximities, to give it a non-exclusive federating content finally worthy of its name? Everything remains to be done, reinvented, brought about, organized, founded, meditated on, thought about ... What could be more exciting for a beginning philosopher?

The end of a cultural cycle, the emergence of another

Steeped to down to the cells in Greek and Latin humanism, contemporary sciences united with ancient languages, I liken my soul to the music of Couperin, Rameau, Ravel and devote my life and tire my flesh at the task of a pure language into which the spirit could descend. More cultivated than me, Stefan Zweig, Georges Duhamel, Romain Rolland, Aldous Huxley or Rabindranath Tagore, before the Second World War, exchanged letters of hope and contributed to founding a few international institutions so that Europe, and subsequently the world, might live in peace. A few months passed, and the worst barbarism was unleashed in the middle of that space of knowledge and humanities in which nonetheless these best among our fathers, intelligent and of good will, trained in this humanism, kept watch. If they ran aground and failed against their will, since one of them even lost his life in this task, this shipwreck forces us to ask what defensive lacks this humanism presupposed, not of course for bringing about these horrors, but for not being able to stop them.

How was this culture, whose loss we lament, not able to prevent Rome or Greece from collapsing with a noise that still resounds in certain ears, or the West, which replaced them, from massacring enslaved and colonized peoples, from exterminating women, the poor, innocent children, plants, animals, that which breathes and doesn't breathe, and, to finish, from destroying this same culture from which it nevertheless derived, formerly and recently, its justification and pride? How did it not save itself? How are we to understand these dramas? Either we explain them by means of their conditions, or we abandon this explanation. Thus barbarity is perpetuated in and by a culture that erects too few truly effective obstacles in front of it. Did the Renaissance flowering stop the French Wars of Religion? Did the ringing proclamation of the rights of man curb colonial expansion and the extermination of tribes, bodies, goods and cultures? Dense with prestigious scientists and men of letters, did the German university restrain the Shoah? Waved like a standard, did the Hellenism of the greatest of its philosophers prevent him from joining the Nazi Party? Did Western humanism halt Nagasaki and Hiroshima, where the nuclear flash governed the Universe? How many intellectuals from my generation covered up the millions of deaths from Stalinism, Maoism and Pol Pot? In the name of what tolerance do we today claim that the strongest democracy is always the best one? Of course, those who opposed and still oppose slaughter did it and do it in the name of this culture by showing that it is of little use if it doesn't educate a cultivated human not to crush anybody under the weight of his culture. This culture, quite the contrary, inspires him to refuse to construct a niche starting from which he would always and everywhere be right; it consists in what is left when one has destroyed this ambition inside oneself.

The old humanism remains, in the form of a memory, on the bank of another antiquity, which the new era is leaving. This book has just drawn the anastomosed river separating the shore on which we are surviving, left behind. It's a question of resurrecting. Of educating this other human that's in the process of being born, of founding a new culture in which obstacles to the return of barbarity increase, of unfolding another grand narrative starting from an encyclopaedia of science, of conceiving a philosophy, of imagining a politics, of building another city, as in the times of Erasmus, Rabelais or Montaigne, after the scholasticism of the Middle Ages, as in the times when Saint Augustine built, on the ruins of the Roman and terrestrial city, the City of God, as in the times when Homer was preparing the poems that founded Greek *Paideia* on the destruction of Troy, as in the times when Jesus Christ paid with his life for the recognition of innocent victims, when Abraham stopped, above Isaac's head, the human sacrifices perpetrated by polytheisms …, not forgetting to take into account an even deeper rupture than the previous ones. What would a civilization be worth that wouldn't recommence, that wouldn't have to recognize, one fine morning, that leaves fall from apoptosis in the autumn and that the immense process of hominescence ineluctably passes through recommencing springtimes?

Of course, very often I no longer recognize my colleagues from the generation that didn't want to hand anything down, or my students, forever orphans of the old humanities, deprived of the knowledge and of the view of the world they offered, disabled of the heart-rending meaning of beauty, deprived, sometimes, of meaning full stop. These contemporaries, who neither read nor write, immersed in the noise of the new oral civilization, do they exist, do they live, do they think better or worse than I do? Who knows this? One more cycle is being completed. We are witnessing another fall of Troy, but I refuse to imitate the black widow, the venomous spider Andromache, whose duty of remembrance keeps watch in the centre of her sticky web and whose memories retained at all costs forbid all life to young people and condemn them to death: a lethal tragedy, whose Muse, precisely the daughter of Memory, possesses a poison for the delaying of culture in the face of the robust health of forgetfulness. I don't miss this burning Troy, whose blackened walls some new Schliemann may rediscover in several centuries; it sometimes looks like Hiroshima. Rereading the end of Book 20 of the *Iliad*, entirely given over to the celebration of Achilles' exploits, who wouldn't throw up before the joy inspired in the hero as well as the poet by the victim clasping his bowels with his hands, the marrow spurting from the spine, the brain that spatters and the liver shooting out of the body amid the dark blood filling the tunic: the naked absence of pity? The scientists I just spoke about happily didn't understand what their ancestors were saying because, precisely, everything was happening in the realm of the dead.

And since we are returning to a first innocence, we might as well rebuild, without these dangerous seeds, for the sake of a newness seeking to minimize major crimes. Culture is recognized less by great men and their inventions than by a reduction, at least as a trend, of the residual remains of the original sin of violence. So let's cheerfully abandon Achilles and Alexander, Horiatii and Curiatii, Roland and Napoleon, illustrious men à la Plutarch, whose glory shone forth from having well and overabundantly killed. The West no longer forms any projects because certain of the projects its culture taught us to celebrate ended with an extermination. Long live the opening of a culture of peace!

Envoi

Can I, lastly and maybe above all, maintain the least trust in the sensibility, the reason, the judgement, the very virtue of these philosophers, ancient and modern, from Plato to Saint Paul and Saint Augustine, from Rousseau and Kant to Schopenhauer, who, all and with a single voice, claimed that women, their companions, obviously their equals, basically amounted lower animals? How could they have looked down to this extent on existence's only real amenity, a courage and stamina always above males, the sweetness of their sleep, the reciprocity of caresses, the mother of their children, she who, amid the linens, welcomes newborns, the sick and the dying? When these men exclude half of it, how are we still to listen to their arrogant chatter regarding humanity? This blunder weightily thrusts them outside thought. What did they know about the love loudly proclaimed in the fine word *philosophy*?

A crisis of consciousness

When its vibrations vanish into the silence that precedes it, follows it and penetrates it the way calcareous water percolates across the stalactites and stalagmites whose points sculpt the details of a cavern, music constructs the consciousness of time. Our own duration traverses us the way a note draws out, a melody flies, the way arpeggios claw their intervals, the way noises and sounds enter and exit silence; these passages leave traces the way a river laden with alluvia gradually builds its own banks; but collapsing by underminings, eddies and swirls, the banks flow the way the water does, although with a different rhythm; it can even be said that, through evaporation and rains, the liquid continuously returns but that the earth flees, even more fluid than the flowing; inside these mixtures as refined as symbioses, one component solidifies and the other one flows along the traces stabilized in this way; but they can exchange roles and functions so that consciousness percolates in time and time in consciousness the way music does in silence and silence within music; just as earth in water and water in earth. The nuptials of music and silence construct the internal cathedral in which consciousness and time are wed together. We even owe clamours and calm more consciousness than we give to speech because speech's meaning inhabits the fluctuating house partly built or dismantled by the clamours and calm.

Every change in listening and hearing, in the syntax of music scores, therefore causes turbulence in this consciousness mixed with or formed by time. To hear music in the past we had to play the violin, the harp, the horn, listen to a pianist passing through, know how to decipher motets or sonatas, rare expertises. Radios, discs, audio systems, portable music players now fill the city and the day with a flood of noise having no islands of silence, so much so that the internal consciousness of time has abruptly changed, at least that of our contemporaries. I no longer regard the works of Bergson, Husserl and Heidegger on this subject as being philosophical but read them with interest as descriptions of a past, one which an ethnology curious to describe how our predecessors experienced the flow of their duration, long ago, will soon deal with.

For a different consciousness appears as soon as silence is lacking, as soon as music, inevitable, captures our own time and expels us, as these Ancients would say, towards an external slavery. For changing durations, consciousnesses and, in total, internal consciousnesses of duration cannot help but cause the self to mutate since time sculpts it, and noise, music and sonorous speech together shape and pilot time. For who am I if not an invariant across variations, a residue of permanence across multiple fluctuations, a repose, a calm, a silence across vibrations? Who am I if not an internal equilibrium stabilized by the inner ear, whose outer ear, its closest neighbour, receives, in the hollow of its pinna, the noisy mobility of the external universe, if not a silent invariance drummed by these variable trumpeted mobilities? The self proceeds from time, but time itself proceeds from the void and the commotion, absolute beginning.

The archaeology of the self and new subjects

Formerly, a slow and long time, *adagio e piano*, constructed an 'I' in the continuous silence pierced with highly rare music or noise and by a sovereign word descending, even rarer still, weak and mortal, among these loud and, in comparison to it, perpetual rumblings. For do you know of a single word that can be drawn out like a note, a single meaning traversing the noise without being aided by sound? Thus, the history of the 'I' began with Saint Augustine and then Descartes, both of them authors of treatises on music, just like Rousseau, a village soothsayer impassioned, him too, by his little *ego*, so as to vanish today into the thundering of radios. Before them, before Christianity, before we sang *(ego) credo*, I believe, before the incarnate Word came, there wasn't any. Yes, the 'I' is defined by an Incarnation of the Word (noise and sound compose its first flesh), an incarnation for which Descartes, alone in his silent stove-warmed room, found a rare variant in the verbal expression of thought.

The universal flood of noise – sounds, music and discourse mixed together, *presto e fortissimo*, erasing the silence – destroys the old agency of the 'I', the way a thin and fragile vase would explode by dint of vibrations, to the profit of a transparency thrown towards the perpetual present, forming an exterior with any interior, weaving relations without reserving any substance for itself, sparkling multiplicities without any nucleus. Formerly a dense seed or dark bit of gravel, single and hard, the self becomes multiple, criss-crossed, mosaic and shimmering.

This is the charm of our grandchildren, closer to Montaigne and La Fontaine, harlequins, than to Descartes and Kant, dark, sad and profound. They no longer have or no longer are the same subjects. What could be worrying about that since such changes have so often adjusted the soul, white and fluid, aquatic and adjustable, possible and contingent? This morning, their innumerable smile succeeds the old cramped confinement.

I bequeath to them the rare music and the silence.

NOTES

Deaths

1 Neighbour = *prochain*, which means one's fellow human, as in 'love thy neighbour'. *Prochain* is etymologically related to *proche*, near, a relation Serres will often play with in this work. I will always translate *prochain* as neighbour, except when it is meant in its adjectival sense of nearness. However, Serres does occasionally make use of the term *voisin*, which means one's spatial neighbour. But I do not believe this will lead to any confusion. (All footnotes belong to the translator.)
2 Newnesses = *nouveautés*, which could equally be translated as innovations. But Serres is often emphasizing the new with this term, and 'innovation' has a very technological ring in English. That said, I do on occasion make use of the latter term as well as 'new things', depending on the context.

Chapter 1

1 If you search for 'chauffe-lit moine' on the internet you should be able to see what Serres is talking about here.
2 Corpulence = *corpulence*, which is probably a play on the French for body, *corps*.
3 Agency = *instance*, probably meant in something akin to the Freudian sense of an agency of the psyche.
4 Train = *m'entraîne*, which, outside a sporting context, can also mean to practice.
5 Entraining = *entraînant*, which is the same word I've been translating as 'to train'.
6 In French, the universal law of gravitation is expressed as the universal law of attraction.

7 Sur-vives = *sur-vit*, which refers to living more clearly than in English. Etymologically, it might mean over-live or super-live. In the following sentence, life is *vie* and survival is *survie*.

8 Serres is referring to the distinction of *natura naturata* and *natura naturans*.

Chapter 2

1 Being-there = *Être-là*, a Heideggerian term usually translated simply as dasein in English. This entire passage revolves around Heidegger's distinction of Being and beings and his idea that Being has been forgotten since the Presocratics.

Chapter 4

1 Old technologies = *anciennes techniques*; new information technologies = *nouvelles technologies*. In this context, *technique* and *technologie* usually mean the same thing. But Serres is distinguishing two kinds of technology, as will be spelled out below. All occurrences of 'technology' in the following subsection refer to information technology. Generally, when the context is clear, I will not modify technology with 'information' to translate *technologie*.

Chapter 6

1 A short story by Alphonse Daudet.
2 The idea that the nectarine is the result of a cross-breeding between the peach and the plum is a widespread, if erroneous, belief in France.
3 Déraciné = *dépaysés*, which means not being in one's *pays* or country, as well as the disorientation that might arise from this. The word *pays* itself derives from the Latin *pagus*, surveyed land, which Serres connects to farming and peasants.
4 *Exploitation*, in French, can mean farming as well as exploitation in the sense of abuse.

5   In 2001 in the UK, over 10 million sheep, cows and pigs were killed and burned to prevent the spread of hoof-and-mouth disease.

6   The latter reference is more than a little cryptic, but I'd guess it has something to do with the death of Princess Diana.

Chapter 7

1   Subdue = *dompter*; tame = *apprivoiser*. Both terms can be translated as to tame. But *dompter* implies harsh methods (as in taming a lion), while *apprivoiser* is gentler (as in taming a stray cat). One possible etymology has *apprivoiser* derive from the Latin *privatus*, so *apprivoiser* can be seen to mean making something private, making it one's own.

2   Fébus wrote the *Livre de chasse* (Book of the Hunt) in 1388. Xenophon wrote *Hunting with Dogs*.

3   Appealing to the older meaning of commodious, namely convenient or suitable. It is entirely possible that, like 'consciousness', Serres sees the *com-* in *commode* as meaning 'with', the mode of the with.

4   How = *comment*, which derives from the Latin *quo modo*.

5   Sur-vivor = *sur-vivant*. *Sur* means over, beyond or even super, and *vivant* means liver or living thing, so *sur-vivant* literally reads as over-liver or over-living thing.

6   Léonce de Lavergne, *Essai sur l'économie rurale de l'Angleterre, de l'Écosse et de l'Irlande* (1854), translated into English as *The Rural Economy of England, Scotland, and Ireland*.

7   Some linguists believe *aîtres, âtre* and *être* are etymologically related.

8   Enclosed farm = *ferme*, which is well named due to its resemblance to the French for closed or closed off, *fermé*.

9   Beauce is one of France's traditional breadbaskets.

10 1 Corinthians 13:7.

Chapter 8

1   A reference to La Fontaine's 'The Milkmaid and Her Pail'.

Chapter 9

1 Joachim du Bellay wrote a poem called *Ô Marâtre Nature* or Oh, Harsh Mother Nature.
2 To prevent confusion, 'law' here is meant in the legal sense.

Chapter 10

1 For those not familiar with France's divisions, this list goes in descending order from towns at the department level to towns at the arrondissement level, and lastly towns at the canton level.

Chapter 11

1 'The steam-powered Human Beast' refers to Zola's novel *La Bête humaine*, in which trains are featured.

Chapter 12

1 As I have already mentioned (see Chapter 4, note 1), *technique* and *technologie* are, in this context, synonyms in French. To deal with the distinction Serres is making in this subsection and the next, I have translated the first term as hard technology and the second one as soft technology. In subsequent sections, I will not translate *technologie* as soft technology except when the context isn't clear.
2 Laptop = *ordinateur portatif*. The French phrase explicitly evokes portability.
3 Lawless place = *lieu de non-droit*. In France, a *zone de non-droit* is a no-go zone, a zone where law is not enforced. But Serres is using the phrase more broadly.
4 The French refer to Robin Hood as *Robin des bois* or Robin of the Woods.
5 *Perdre* usually means to lose but can also mean to leak.

6   The French words refer to thumbs, elbows and arms more obviously than their English counterparts do.

Chapter 13

1   Dismissal of charges due to no grounds for prosecution = *ordonnance de non-lieu*, which literally reads as ruling or order of non-place.
2   Flaubert, it is said, would declaim every word he wrote in a loud voice as a test.

Chapter 14

1   The ferret of the woods refers to a children's game similar to 'Hunt the Slipper'. The children play it while singing about the ferret of the woods who runs: the ferret, it's running, it's running; the ferret of the woods, my ladies.…

Chapter 15

1   The quote is from Victor Hugo: *Feuilles d'automne*, 'This Century Was Two Years Old.'
2   Conscience = *consciens*, which can also mean consciousness.
3   *Adolphe* and *Dominique* are novels by Benjamin Constant and Eugène Fromentin, respectively.
4   Tense = *temps*, which also means time.

Peace

1   In French, a Pulcinella's secret is an open secret.
2   *Captain Fracasse* is an 1863 novel by Théophile Gautier. The name Tranche-Montagne has come to mean a braggart.